The Logging Business Management Handbook

Also published by Miller Freeman Publications for the forest products industry:

Logging Practices
revised edition
by Steve Conway

Timber Cutting Practices
third edition
by Steve Conway

Dry Land Log Handling and Sorting
Planning, Construction, and Operation of Log Yards
by Charles M. Hampton

Quality Control in Lumber Manufacturing
edited by Terence D. Brown

Small Log Sawmills
Profitable Product Selection, Process Design and Operation
by Ed M. Williston

Lumber Manufacturing
Design and Operation of Sawmills and Planer Mills
by Ed M. Williston

Plywood Manufacturing Practices
second edition
by Richard F. Baldwin

Modern Particleboard and Dry-Process Fiberboard Manufacturing
by Thomas M. Maloney

Saws
Design, Selection, Operation, Maintenance
by Ed M. Williston

Modern Sawmill & Panel Techniques
Volumes 1 and 2
Proceedings of the North American Sawmill & Panel Clinics, Portland, Oregon, 1980 and 1981

Modern Sawmill Techniques
Volumes 1 through 9
Proceedings of the North American Sawmill Clinics

Modern Plywood Techniques
Volumes 1 through 7
Proceedings of the North American Plywood Clinics

Electronics in the Sawmill
Proceedings of the Electronics Workshop at the Sawmill & Plywood Clinic, Portland, Oregon, 1979

Business Management for Sawmill Operators
Proceedings of the Business Management Clinic for Sawmill Operators at the Sawmill & Plywood Clinic, Portland, Oregon, 1979

Sawmill Techniques for Southeast Asia
Volumes 1 and 2
Proceedings of the Southeast Asia Sawmill Seminars

THE LOGGING BUSINESS MANAGEMENT HANDBOOK

RONALD R. MACKLIN

A FOREST INDUSTRIES BOOK

MILLER FREEMAN PUBLICATIONS, INC.
San Francisco

DEDICATION

This book is dedicated to my wife, Carmen Lynne, without whose advice, love, and constant encouragement this manuscript would never have been completed.

This book is designed to provide the author's opinion on the subject matter covered, based on the data available to him at the time of publication. It is sold with the understanding that neither the author nor the publisher is engaged in rendering legal, accounting or other professional services. If legal advice or other expert assistance is required, the services of a competent professional should be sought. To the best of the author's knowledge the information contained in this book is reliable and complete, but no warranty is made in this regard. All expressions of opinion are subject to change without notice. The author and publisher expressly disclaim any liability for any risk, loss or damage incurred as a consequence of the use and application, either directly or indirectly, of any advice or information presented herein.

Library of Congress Catalog Card Number: 82-083344
International Standard Book Number: 0-87930-146-5

Printed in the United States of America.

SHORT CONTENTS

DETAILED CONTENTS

SECTION ONE
Organizational Structure and Financial Planning

1 Structuring Your Business

2 Money Sources

SECTION TWO
Operations Management and Controls

7 Simple Aids in Cost Accounting

8 Establishing Controls and a Cost Analysis System

9 The People Part of Your Business

10 The Equipment Part of Your Business

11 Maintenance of the Entire System

LIST OF FORMS

PREFACE

During my early exposure to the business of logging, a friend told me a story about an old-time logger. The story is so true and accurate that I must share it with you.

It seems this old logger, after years and years of hard work in the woods, suddenly came into an inheritance of over a million dollars. On being interviewed about his future plans, the old logger was asked, "What will you do with your life now that you have over a million dollars?" The old man reflected for a moment and then with an air of eager anticipation he said, "I think I'll just keep on logging till it's gone."

Most of the loggers I know love what they are doing as much as that old timer. They love it so much that they often lose their perspective when it comes to making a success of their business.

Everyone in the logging business must periodically ask himself several important questions. For example:

- Am I losing money logging?
- Why am I in the logging business?
- How much do I have tied up in my business?
- If I am losing money, what can I do about changing the situation?
- How can I make a fair return on my investment?

In this book I'm going to share some things with you that could make the difference between staying in business or going "belly-up." The information will allow you to determine whether you are working to break even or working for a profit. The basic ideas and principles outlined in this book can be applied to almost any small business.

I don't pretend to know everything about the logging business, but what I do know has been learned in 5 of the hardest years I have ever worked. That work, coupled with 20 years of management and administrative experience, gives me a small "leg up" on being able to talk about the business side of logging.

The information in this book is designed for loggers—those who want to go out on their own and those who already have a business, whether it's successful, just breaking even or in need of some new ideas.

The book has two sections. The first—Organizational Structure and Financial Planning—explains how to structure a business based on time-tested principles of organization, where to look for the money your company needs and how to get it, and how a company grows (and the pitfalls to avoid). It also includes essential money management information to help you plan your company's future: the effect of the new depreciation rules on your company's balance sheet, how to determine whether investment in new equipment is justified and whether to buy or lease it. And, there are down-to-earth suggestions for turning your business around if it's in trouble.

Section Two—Operations Management and Controls—explains step by step how to set up a cost analysis system that will tell you exactly the condition of your company. The system also gives you all the information you need to turn a marginal operation into one that shows a profit. This section also puts together information on important aspects of management-employee relations—among them, salaries, equal employment opportunity regulations, incentives and fringe benefits. It also contains tips on getting the most out of your equipment—the core of your business. Finally, a series of cost-analysis and personnel management forms are included. You can modify them to meet your needs or use them as they are.

ACKNOWLEDGEMENTS

To survive in the world of business, one requires the cooperation and understanding of many people. During the course of these relationships, one gets advice, encouragement and help from many of these individuals. I wish to thank the following friends who helped me to live the concepts put forward in this book: Gary Parrish, financial analyst; Lee McLaughlin, collection supervisor; Ken Campbell, fuel supplier; Doug Dick, logging supplier; Pete Carson, loan officer; Ike Kingrey, insurance specialist; Tom Griggs, CPA; James L. Hershner, attorney at law; James W. Grant, mechanical engineer; Elvin A. Carter, contract logger; F.F. (Monte) Montgomery, President, Associated Oregon Loggers; James A. Hawkins, sales manager; and James Luke, entrepreneur.

Thanks are also given to Bev DeWitt for her excellent suggestions and editing of the manuscript and to the staff of Miller Freeman Publications, including Jann Donnenwirth and Anu Prints-Seldin, for their guidance and help in the publication of this book.

Ronald R. Macklin
Lowell, Oregon
September 1982

SECTION ONE
Organizational Structure and Financial Planning

1 Structuring Your Business

2 Money Sources

3 The Evolution of a Logging Business

4 Understanding Depreciation and the Investment Tax Credit

5 Evaluating Equipment Acquisitions

6 The Business Turnaround

CHAPTER 1

Structuring Your Business

WAYS TO STRUCTURE A BUSINESS

Do I want to operate my business as a *partnership,* a *sole proprietorship,* or a *corporation?* This is a question that many independent loggers ask themselves when they decide to go into business for themselves or if they are considering a change in the way they've been doing business. In order to make this decision, you must know something about how each type of business works and the advantages and disadvantages of each. This information is given below in layman's terms to eliminate as much confusion as possible.

Proprietorships and Partnerships

Advantages. Proprietorships and partnerships—two fundamental ways of running a business—are slightly less expensive to start than a corporation, since there are no incorporating fees. They are also a bit less formal. Here are the advantages:

- Organization costs less since there are no incorporation fees. (It is advisable, however, to consult an attorney when organizing it.)
- Record keeping is less formal than for a corporation.
- Owners file one tax return for the business.
- Owners can deduct losses that might be incurred during the early life of a business from other personal income.
- The limit of tax-deductible contributions to Keogh type pension and profit-sharing plans has been increased. This has reduced the tax advantage of benefit plans previously available to a corporation.
- Profits of a partnership, unlike dividends paid by a corporation, are not subject to a second federal income tax when distributed to the owners. However, whether or not this is an advantage, taxwise, depends on certain other factors, namely:

  The individual tax brackets of the owners compared with the tax bracket of the corporation.

  The extent to which double taxation of earnings is eliminated by the ability of the corporation to deduct the salaries paid to owners and retain earnings in surplus.

Disadvantages. Some of the main disadvantages of proprietorships and partnerships are:

- Unlimited personal liability. The owners are personally liable for all debts and judgments against the business, including liability in case of failure or other disaster.
- In a partnership, each member can bind the other(s) so that one partner can cause the other(s) to be personally liable.
- All profits are personally taxable to the owners at rates that are higher than corporate rates.
- If the owner(s) dies or becomes incapacitated, the business often comes to a standstill.
- The owner(s) does not have the full tax benefits of the tax-deductible plans (including pension and profit sharing) that are available to a corporation.

Incorporation

Advantages. There are a considerable number of advantages to incorporating. They include:

- The personal liability of the founders is limited to the amount of money put into the corporation (with the exception of unpaid taxes).
- If a business owner wishes to raise capital, a corporation is more attractive to investors, who can purchase shares of stock in it.
- A corporation does not pay tax on monies it receives in exchange for its stock.
- There are many more tax options available to corporations than to the other forms of doing business. Owners can set up pension, profit-sharing and stock-option plans that are favorable to themselves.
- A corporation can be continued more easily in the event of the death of its owners or principals.
- Shares of a corporation can be easily distributed to family members.
- The owners (stockholders) of a corporation that is discontinued due to a lack of success can, in the year the corporation is discontinued, have all the advantages of being incorporated, yet be able to deduct up to $25,000 on an individual tax return or $50,000 on a joint return of money invested in the corporation from personal income.
- The owner(s) (stockholders) of a corporation can operate with all the advantages of a corporation, yet be taxed at personal income tax rates if this option provides a tax advantage.
- Owner(s) can quickly transfer their ownership interest, represented by shares of stock, without the need to dissolve the corporation.
- The corporation's capital can be expanded by issuing and selling additional shares of stock.
- Shares of stock can be used for estate and family planning.
- The corporation can ease the tax burden of its stockholders by accumulating its earnings, provided that the accumulation is not unreasonable and is for a business purpose.

- A corporation is a legal "being" separate and apart from its owner(s) (stockholders). It can sue and be sued and can enter into contracts.
- A corporation may own shares in another corporation and receive dividends, 85% of which are tax free (with some exceptions).

Disadvantages. The following disadvantages should be considered carefully before deciding to incorporate:

- The owners of a corporation file two tax returns, individual and corporate. This may require added time and accounting expense.
- Unless the net taxable income of a business is substantial—i.e., $25,000 or more—there may be no tax advantages to incorporation.
- Maintaining the corporate records may require added time.
- If debt financing (i.e., a loan from a bank) is obtained by a corporation, the fund source may require a personal guarantee by the owner(s), thereby eliminating the limited liability advantage of a corporation, at least to the extent of the loan.
- The biggest single disadvantage to incorporating is the high initial cost.

STARTING A NEW BUSINESS

After you have decided on the form in which you will do business but before actually conducting business and putting people to work, you must accomplish the following things:

- It is highly recommended that you consult an attorney and an accountant prior to putting your organization together.
- If you are not incorporating, it is necessary, in many states, to register your assumed business name with the corporation commission. You can call your state offices and the forms will be mailed to you.
- Obtain a Federal Identification Number, preferably before you make your first payroll. Every sole proprietor, partnership, corporation or other employing entity must obtain a Federal Identification Number. Use Federal Form No. SS-4. This number covers federal income taxes, federal unemployment taxes, and Social Security withholdings.
- Also obtain a state employer account number for state withholdings, preferably before issuing your first payroll. This number will take care of your state tax withholdings and other payroll withholdings.
- Obtain worker's compensation coverage. *Make sure you do this prior to sending employees to the work place.* Every state requires an employer to provide worker's compensation insurance for all employees. Coverage can be obtained through commercial companies or associations, some states have state-sponsored plans or you can be self-insured if you have the resources available to you. Regardless of how you get the coverage, you must register with the worker's compensation department in your state.
- Obtain and read the current summary of wage and hour laws for both state

and federal requirements. These are usually available through your local bureau of labor.
- Review the summary of Equal Employment Opportunity laws, both state and federal. This document is also provided by your local bureau of labor.
- Obtain copies of your state's current laws regarding safety and health codes. There will usually be a specific code for logging. You must be in compliance with the applicable laws in your state.
- Make sure that you comply with your state's laws regarding independent contractor status. You must be recognized by your state as an independent contractor in terms of payroll withholdings, insurance coverage and other important areas. In many states, failure to comply with the local requirements regarding this status can cause much confusion between you and your contractors and subcontractors.

If you follow this checklist, you will have laid a solid foundation for conducting your business. There are many other things involved in conducting a logging business, but since each state varies in its practices, we cannot adequately cover every aspect of setting up your business. For example, you will have county taxes, license fees, PUC requirements, federal fuel consumption reports, federal business reports and many other things that are governed by the size of your business and the state in which you operate.

ESTABLISHING AN ORGANIZATION

The basis for the smooth evolution of any small business into a larger business is establishing an organization. This will create an environment in which both your business and your people can grow. Your organization can be as complicated or as simple as you want it to be. Before we get into the mechanics of structuring an organization, let's talk about management and why it is important.

The Purpose of Good Management

According to today's research into small business failures, it has been proven beyond any reasonable doubt that bad management is the primary contributing factor. Are you a good manager? Let's explore that question for a moment.

First, what is management? If you can't answer that question, you can't possibly be a good manager. Management is people; organized responsibility; the translation of knowledge into positive results; the ability to plan, organize, motivate, coordinate and control. So a good manager sets objectives, organizes, creates a team by stimulating his people to action, communicates, analyzes, appraises, reviews and interprets performance. All this action is to create an enterprise that offers your customers an advantage—one that will gain you preference when they choose a contractor. You will build a management strategy around this advantage, which will create a niche for your company so that it can withstand any competition and the ravages of today's uncertain economy. This management strategy is your key to long-term survival.

Functions and Relationships

To manage effectively, a manager must develop the kind of organization that will carry out his management strategy. In organizing, the manager is concerned with the following three activities through which he exerts his organizing function:

- He must divide the work of his company so that the work being done by one person is not duplicated or even overlapped by another.
- He must become the source of authority. Unless the manager is the directing authority, he will be operating a mob rather than a company.
- He must establish among his people orderly working relationships that result in a minimum of human friction and a maximum of human effectiveness.

Let's take a moment to touch on these points in a little more detail. They are very important to the success of your organization. When a manager organizes the work that needs to be done to reach his objectives, he must divide the work of his company into manageable parcels that can be handled by individual persons. Each such "work package" is made up of closely related items of work that best fit the knowledge, skills and ability of the person who will do the work.

The next job in organizing is to see that each person knows exactly what he is expected to do and to whom he should turn for direction when needed. This means that the worker must know what he can and cannot do, and exactly how far he can go before he has to go to another for approval.

Finally, the manager needs to set up the proper relationships among the members of his working team. It is clear, therefore, that he must establish rules for teamwork so that the efforts of the team members blend together toward clearly understood objectives.

To make your organization a reality, you should first know the pertinent facts about each aspect of organization so that you can apply the principles of collective effort to your company. The aspects of organization will be covered in the following order:

- The four principles of organization.
- The four basic types of organization.
- The six-step procedure for reorganization.

THE FOUR PRINCIPLES OF ORGANIZATION

The four principles of organization, which are the basis of any good organization, are:

- Unity of command.
- Span of control.
- Homogeneous assignments.
- Delegation of responsibility with corresponding authority.

These principles are based on human behavior as it shows itself in group working relations. They have been developed over centuries of human experience, and are as basic as the principles of modern science. Like the laws of science, the principles of organization cannot be ignored without drastic consequences.

The four basic principles of organization are not "rules" as we know them in sports, where the rules dictate exact decisions for specific actions on the part of the players. Rules such as those can be changed frequently. The principles of organization, however, are fundamental, basic, invariable laws of human behavior. Regardless of the size of your organization, the four basic principles of good organization must be applied to each and every one of the people involved in your company, or some degree of failure will always result.

Unity of Command

In effect, unity of command states that "no person shall have more than one boss." It does not mean that a superior cannot have several subordinates. It does mean that no employee can work effectively for more than one immediate superior at one time. A worker who is subject to the conflicting orders of several superiors is likely to be inefficient, confused and eventually irresponsible. Of the four principles of organization, the principle of unity of command is the one most often violated.

To ensure the effective operation of the principle of unity of command, you must take the following steps:

- Prominently display a chart of your organization and explain it to everyone concerned. This will make the proper channels clear to everyone.
- Instruct your subordinates that they are to accept orders from no one but their immediate boss. Anyone else attempting to give them orders should be referred to their immediate boss.
- Obtain the cooperation of the person violating the principle of unity of command by explaining the harm he can cause. Get him to see that if such violations continue, he will destroy all authority for everyone.
- Carefully define the jobs of your subordinates.

When you get everyone to stay within the chain of command, you will take a giant step forward in maintaining authority. This will do more for efficiency than any other single thing you can bring about.

Span of Control

The second principle of organization requires that the proper span of control be given every supervisor within your company so that each supervisor has the correct number of subordinates. This principle has three aspects: (1) people, (2) distance and (3) time.

The People Aspect. The number of subordinates any supervisor can efficiently control depends primarily on the duties and abilities of the subordinates. The span of control will vary according to the jobs performed and the people performing them. Perhaps one man can direct the work of 15 or 20 factory workers on an assembly line. But if the work is diversified, highly technical or otherwise complicated, one man may have difficulty supervising as few as 4 or 5 people. This is definitely the case in logging. So don't expect one man to spread himself too far.

The Distance Aspect. If a supervisor's subordinates work within a short distance of him, his problems of control are reduced. On the other hand, if his subordinates are spread throughout the building or the countryside, the number of men he can supervise at one time automatically decreases. If he tries to handle a large group despite the increased distance, one or two workers located near his base of operations will receive most of the supervision. Those located at a great distance will receive only rare attention.

The Time Aspect. There will always be time limits for doing any work, and working with these limits is one of the objectives of any company. The supervisor is responsible for setting up his organization in such a manner that it will be able to accomplish the required work within the time limits set. Short-term jobs are easier to supervise than those of long duration because they require fewer controls. So, we can see that time limits are necessary for the proper span of control.

Homogeneous Assignments

In this principle of organization, the word homogeneous refers to the likeness or similarity of the functions performed by an individual employee or group.

Homogeneity can be carried to two extremes. At one extreme is the job that requires a worker to repeat the same operation over and over for a full working day, without relief. Excellent examples of this are setting chokers, running a yarder and running a loader. What can happen is that the work can become so monotonous that the men, accustomed to doing the job mechanically, allow their attention to wander. The results can be injuries and disgruntled employees.

At the other extreme is assigning too many jobs of a miscellaneous nature to a single worker. A good example of this might be a "chaser" on the landing. He is expected to release the chokers on each turn, limb the logs, paint the logs, brand the logs and watch out for his own safety at the same time.

Experts are not in complete agreement as to how many tasks make up a homogeneous assignment, but all agree that there should be no less than 4 and no more than 12 kinds of tasks. This range assures enough variety to prevent boredom yet few enough tasks so that the worker can perform each of them competently.

Most workers take pride in excelling in their field. If the scope of a worker's job is so wide that he is deprived of this opportunity, the result will be dissatisfaction with his job. Too often, employers fail to realize the importance of fulfilling the needs of the employee through job satisfaction. At the same time, too much of a good thing can be equally dangerous.

Homogeneity of work is the spring from which job satisfaction and job pride flow. More than anything else, employee pride and satisfaction cannot be overlooked by any company.

Delegation of Responsibility with Corresponding Authority

In the fourth principle of organization, a supervisor's authority over his unit is parceled out among his subordinates. This is understood to be delegating his authority. This does not mean that the supervisor is relieved of responsibility for overall performance. It does mean that the subordinate is supposed to represent the will of his superior in carrying out the assigned task. When you delegate authority, you should remember that you are not giving it away. If the one you delegate it to falls on his nose, you must take the responsibility for his failure. On the other hand, if the employee does an outstanding job, you should be willing to give him full credit for his accomplishment.

In delegating responsibility, it is mandatory that corresponding authority also be delegated. If this is not done, the boss makes it impossible for the subordinate to get ahead with the work assigned to him.

THE FOUR BASIC TYPES OF ORGANIZATION

There are four basic types of organization: direct-line, functional, line-staff and project management. Each is briefly described but only those types that lend themselves to the logging industry and to small companies in general, are detailed.

Direct-Line

In direct-line organization, authority flows vertically in one continuous, direct line from the top of the company through each of the levels to the lowest position. This is the simplest form of organization. You might recognize it from your experience in the military, because it is often called the "military form" of organization. You can see a simple illustration of this form of organization in Figure 1. This type of

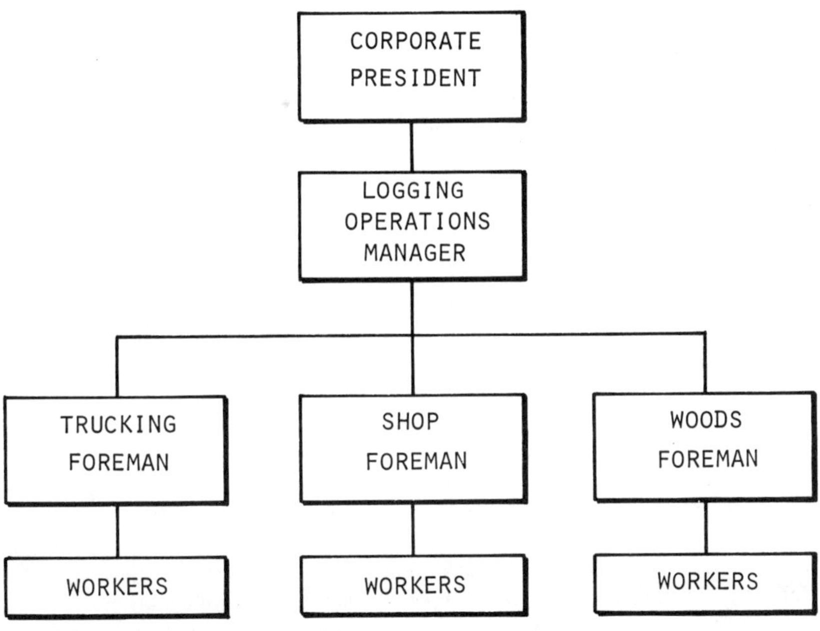

Figure 1. Direct-line organization chart

organization is rarely used in the business world in its pure form, except in very small companies. It can be adapted to a small logging company and is probably the easiest form of organization to use. The main advantage of this type of organization lies in the easy identification of duties, responsibilities and authorities. The main difficulty with this system is that it fails to handle the interrelationships of the various management functions and the increasing inability of one man to achieve mastery of all aspects of management. There is also a tendency within this setup toward empire building.

Functional

A functional organization is based on the concept of a group of specialists at the top level. Each of these specialists has direct supervision over the operating employees who perform those activities within his specialized function. This concept of functional organization originated with Frederick W. Taylor (1856–1915), who outlined the concepts of scientific management and work efficiency. This form of organization has capitalized on the development of special skills and on the separation of manual and mental work. Figure 2 illustrates this type of organization.

The advantages of functional organization are:

- One man devotes his entire time to one task.
- Organizational growth is not hampered, because each man grows in his own specialty.
- Each man gives the company his best effort because he develops according to his best abilities.

Functional organization also has its disadvantages:

- It is confusing from the standpoint of control procedures.
- It is conducive to overlapping authority and to indefiniteness of fixed lines of responsibility.
- It does not permit easy coordination of work, and it depends primarily upon brilliant leadership.

This form of organization is rarely found today in its pure form, although certain features of this system are observed in many companies. It can be found in some logging companies.

Line-Staff

Line-staff is the most frequently used type of organization in large companies today. In its pure form, line-staff has one top position from which the direct line of control runs to the operating components. Since the head of a large company cannot be expected to act as his own advisor in all matters, especially those of a specialized and technical nature, this top position is supplied with a staff of experts in various phases of activities. When advised by his staff, the head of the company issues all orders and instruction downward through a direct line of control. He also

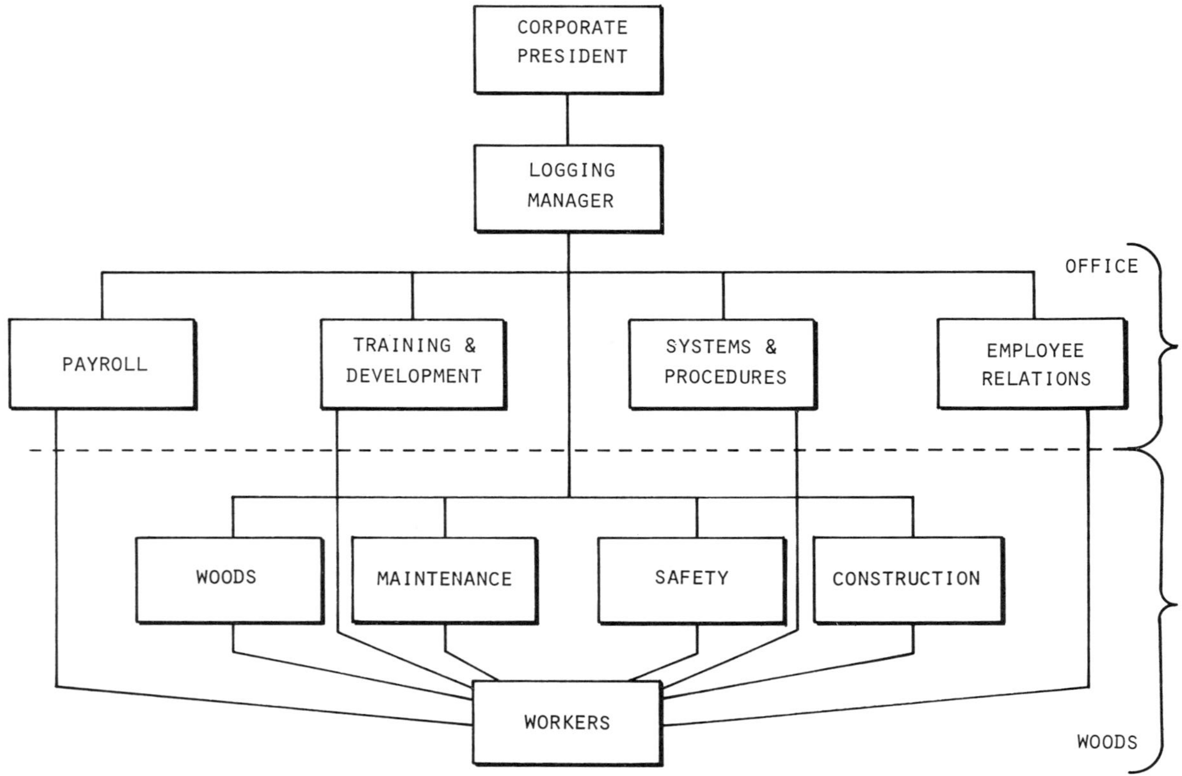

Figure 2. Functional organization chart

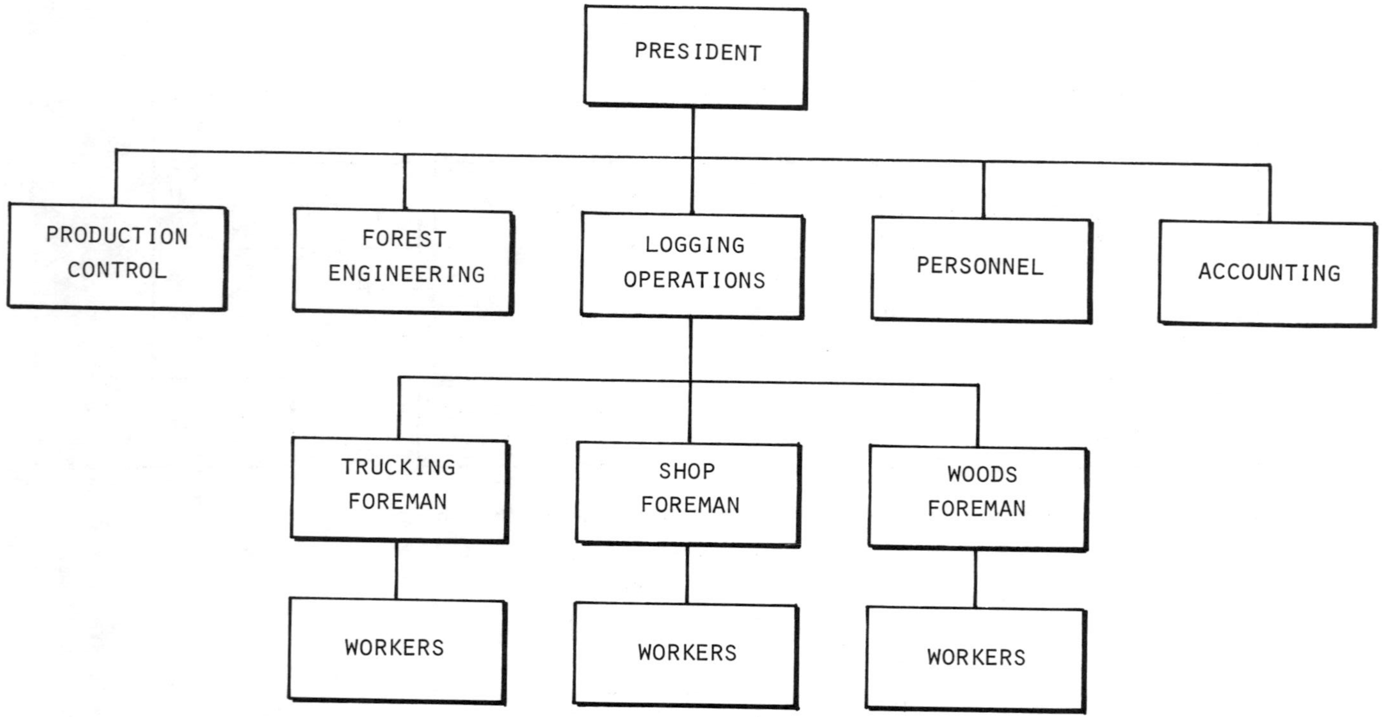

Figure 3. Line-staff organization chart

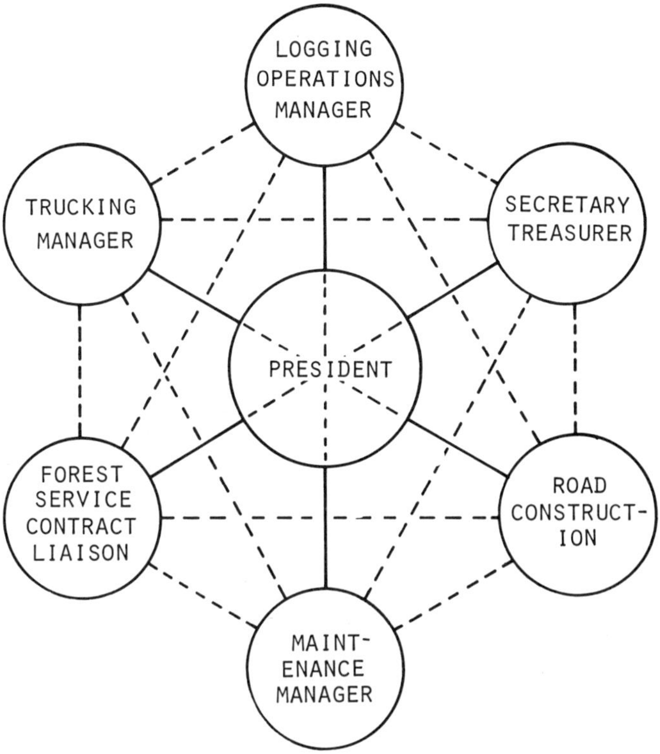

Figure 4. Nonconventional line-staff organizational structure

supervises the execution of these orders by the leaders of his operating groups. You will find two illustrations showing how this type of organization works. In Figure 3 you will see a typical line-staff organization, and in Figure 4 you will see a special adaptation of this form of organization that was used successfully, over a 4-year period, in a logging corporation. This form of organization permits help and guidance to be given to the workers by the specialists, and at the same time it ensures that an individual has only one person to whom he reports. You will notice too that an interface is encouraged between members of the line staff as well as between the head of the company and each member of the line staff.

Project Management

Project management is a specialty type of organization used when a line-staff organization is faced with the problem of completing jobs or projects that are unique and that will only be done once. Since this type of organization is not particularly suited to logging, we will not elaborate on it. Figure 5 provides an example of this type of organization for your information.

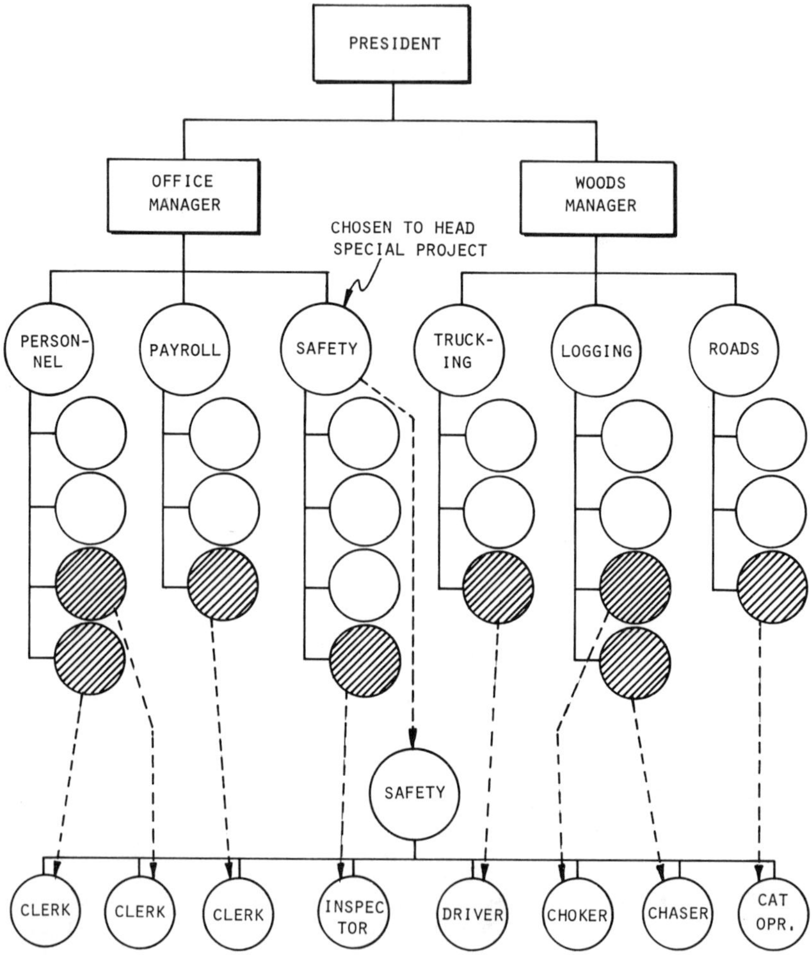

Figure 5. Project management organization chart

THE SIX-STEP PROCEDURE FOR REORGANIZATION

If you currently have an organizational structure but you wish to reorganize, the following six things should be taken into consideration prior to reorganization:

- Determine the functions within your company by making a list of all the jobs performed by each person within each area of responsibility.
- Group the jobs into related or homogeneous areas.
- Determine the number of positions required, both in supervision and in hourly.
- Set up the organizational structure that you want in chart form.

- Allocate personnel to each of the jobs you have listed.
- Now move the personnel to the positions assigned.

Obviously, if you are setting up a new organization, your task in allocating personnel is easier than if you are reorganizing a company and must use personnel already working.

Many times a good reorganization can save a faltering company. Give it serious thought if you have an organization that is in trouble.

OWNER ATTITUDE

Your attitude as the owner of the company is vitally important. For example, let's say you have decided to make organizational changes in your company. Along with these changes you are going to have job descriptions written, a company policy manual written, a company safety manual written, and all kinds of new and useful programs established within your company. How far do you think all this will go if you as the owner are not 100% behind everything that is written and distributed to your employees? If you do not come out in writing in support of all these things, they will not succeed. Once you decide to do these things, stick with those decisions all the way down the line. If you put a man in charge of something, support that man and stand behind him in his decisions. If you make a rule and someone breaks it, stand behind that rule and administer discipline and levy penalties accordingly. If you have a written procedure for doing a job and someone decides to ignore it without a valid reason, stand behind your decision. Your absolute support of everything you do is the only way you will be successful as a leader and as a businessman.

Policy Statement

Here is an example of an owner policy statement. Make sure that all your employees receive a copy of the one you prepare for your company.

> Welcome to XYZ Corporation
>
> As an equal opportunity employer, we hope our working relationship with you will be long and profitable for both you and the company. We have prepared this statement and the attached handbook for your information.
>
> We hope to answer your questions about the company, so you will know what to expect from us and what we expect of you. We hope this will eliminate any misunderstandings that might occur during your tenure with our company.
>
> Should you have questions about any aspect of your job, or about the company, that are not covered in this manual, please don't hesitate to ask your supervisor, one of the office staff, or me.
>
> Signed by you, the owner.

It should also be made clear in the preface to your manual that the company reserves the right to modify, change or delete any of the policies or procedures set down at any time. And that any such modification, change, or deletion will be given to the employees in writing at least one week prior to the effective date of such modification, change or deletion. This added statement makes it clear that the company is not dogmatic or inflexible in its viewpoint and that you as owner have the final say as to what is going to become 'gospel' within the framework of legal and safe operation.

COMPANY IDENTITY

Several years ago, after joining a company as secretary/treasurer and making some sweeping changes in company policy and in the way employees were dealt with, I had a supervisor come to me and say, "You know, Ron, for the first time in 13 years, I feel like I am part of a company and not just a group of guys who go out and get dirty every day."

One of the company policies that we worked on was *identity*. In order for an employee to feel part of something, there has to be something tangible to be a part of. Identity not only helps the employee and the community, but it also helps the company.

Importance to You as Owner

Whether your company is tiny or large, it needs to be identifiable—identifiable by its employees, its customers, its lending institutions and the general public.

Figure 6. Logos of large companies

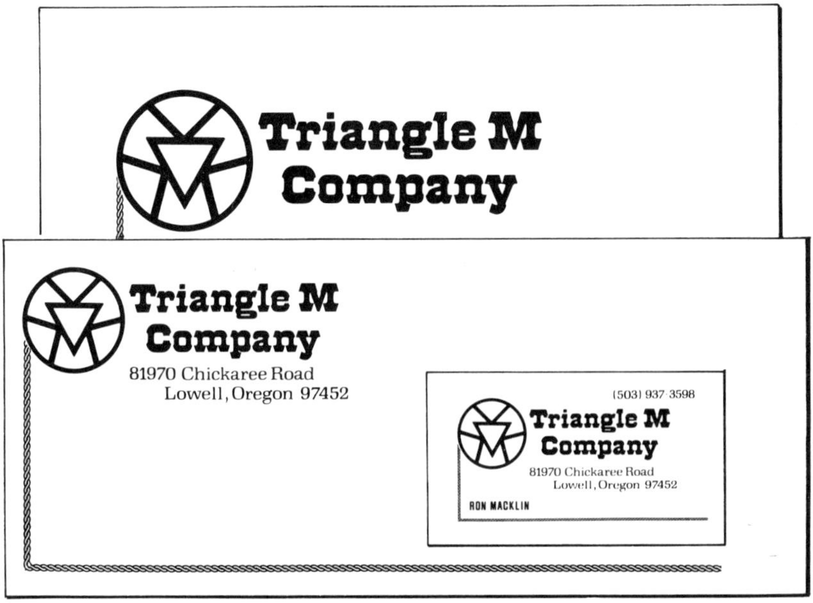

Figure 7. Use of company logo on various items

Corporate identity can be a valuable asset to you as a businessman. Large corporations are recognized by their distinctive logos or trademarks. In Figure 6 you will undoubtedly recognize some of the major symbols used by large companies.

The larger the company, the more important an image becomes. Many large companies have a manager of corporate identity whose only job is to tell the employees what they can use the company logo on and how it should be done.

We are not recommending any such grandiose scheme for your company identity program. We are recommending a small-scale program that will benefit you in several ways.

Methods of Establishing an Identity Program

First, develop business cards, stationery and envelopes for use in your everyday business. If you already have a logo or corporate symbol, you can incorporate it into these materials. Figure 7 shows the various ways in which I use the company logo in conducting my business.

You can carry your identity on to your equipment, buildings and whatever you wish. Perhaps you have all your equipment painted a particular color. It is good to have company colors if it is easy to accomplish and you think it worthwhile. As the equipment needs repainting, you can make sure you use the same colors each time. As time passes, all your equipment will eventually be the same colors.

Regardless of the size of your company, a program of identity is a worthwhile thing.

SUMMARY

Once you have gone through the exercise of doing everything to satisfy the bureaucracy and structuring your organization to operate the way you wish it to, the success or failure of your organization depends upon you. *You* will determine the final results.

Make sure you give the system a chance to work before you decide to change it in any way. Make sure you give the system your energetic support and expect your employees to do the same.

Nothing worthwhile is easy. But you will find the positive results well worth the effort it takes to make it work.

Remember: *results through unity of purpose.*

REFERENCES

Drucker, Peter F. 1980. *Managing in Turbulent Times.* New York: Harper & Row.

Levinson, Harry. 1973. *The Great Jackass Fallacy.* Cambridge, Mass.: Harvard University Press.

Longnecker, Justin G. 1969. *Principles of Management and Organizational Behavior.* Columbus, Ohio: Merrill.

CHAPTER 2

Money Sources

Next to methodical record keeping and accurate accounting systems, which will be discussed in Chapter 7, the ability to manage money and money sources is high on the list of things necessary to go into and stay in business. If you have no talent in this area, the other items in this management system won't work as they should to keep your company strong.

Please don't misunderstand me. You do not have to be a financial wizard to survive. But the more you know about the financial position of your company and how to communicate that to others, the stronger your company position will be in negotiations for money and equipment.

There seems to be a large gap between bankers and businessmen. The businessman generally does not understand the bewildering financial world of the banker, and the banker is amazed at how little the businessman knows about finance. In fact, of all aspects of his business, the businessman usually knows the least about finance.

Before getting into the details of how to get money for your company, let's cover some basic things that you should be familiar with.

KINDS OF CAPITAL

Remember the difference between *debt capital* loans—that is, money you borrow and pay back—and *equity financing*—that is, money given to you in return for part of your business.

Briefly, the difference is this: when you are given equity money, also called seed capital and injection capital, you also must give something back—specifically, a chunk of your business. When you borrow debt money, you are loaned the money and are expected to pay it back, plus interest, in a specific period of time. But, you keep all of your business for yourself.

The obvious advantage of equity capital is that you don't have to pay the money back, and if you are the majority stockholder, you don't even have to declare dividends to the people who gave you the money in return for your stock.

The major disadvantage of equity capital is that you have given away part of the business in return for the money, and so the business is no longer completely yours. This means that you must constantly deal with the wants, needs, and business opinions of others. If you own less than 50% of the business, there is always the possibility that the other stockholders might get their heads together and boot you out of your own company.

Debt capital is borrowed with the intention of repayment, and you remain the sole owner of the business. Of course, if you default on the loan, the bank can claim your assets for liquidation. The advantage of debt capital is that you retain ownership and you get leverage.

The disadvantage of debt capital is that you have to pay the money back plus interest. If you don't, you could lose your business.

ADVANTAGES OF BORROWING

You should avoid excessive borrowing, of course, but you should also be informed of the advantages and benefits of borrowing:

- Borrowed funds can earn more than they cost. The rate of interest that lenders charge is generally lower than the rate of return you expect your investment to earn. As a result, borrowed funds, if used successfully, increase the return on investment over what it would have been without the borrowing.
- Interest payments on borrowed funds are tax deductible. Dividend payments or other returns to investors are not. These facts should be taken into account, along with other considerations, in deciding on borrowing policies.
- Some credit is a matter of convenience. Buying merchandise on open account and services on an accrual basis are examples of credit-for-convenience. Wages are a specific example. Even a small company would find it difficult to make daily wage settlements.

SOURCES OF MONEY

The major sources of debt capital for businessmen are banks, private financial institutions and the government. Each of these categories is summarized below.

Banks

Banks are generally the best source to approach for all types of loans because:

- Banks usually charge lower interest rates than finance companies.
- The banker can give you expert business and financial guidance since the banker is in the business of dealing with other businesses' finances.
- The bank relationship will help establish a good basis for credit references.
- A banker can give you access to different forms of loans.

Private Financial Institutions

Private financial institutions are more likely to be in a position to help the small businessman because they will make loans more readily to "risky" businesses. When your business is new or your credit isn't sterling, a private financial institution will take a chance on you when a bank won't.

Government Financing

What is the advantage of government financing? The primary advantages are good terms, low interest, and low equity requirements (net worth of present or personal cash injection into a new business)—and that's a lot of advantages. The disadvantages are that there are special criteria a businessman must meet to be considered, and the loan proposal is usually more detailed and more difficult to put together than that for other financing.

HOW DO YOU RATE AS A BORROWER?

The lender—who more often than not is a banker—needs certain information on which to base a loan decision. Part of it will be furnished by you, the borrower, and the rest will come from the banker's own credit files and from outside sources. This information is related to what credit men call the "Cs" of credit: *character, capacity, capital, collateral, circumstances* and *coverage* (insurance). Let's review each one briefly.

Character

To your banker, character means two things in particular: (1) The borrower will do everything in his power to conserve his business assets and so ensure repayment of his loan. He will manage his business to the best of his ability. He will not squander his own or other funds. (2) The borrower (you) is a man of his word. When he says that he will repay his borrowing promptly, he means it. If he does not keep his promise, he at least will have made every possible effort to do so.

Capacity

The management skill shown by the small-businessman in using his investment and enlarging it is another important business asset. For those just embarking on a business career or entering a new field, however, past performance—however successful—does not qualify a person to direct all the activities involved in operating a different business.

Capital

The small-businessman's investment in his own business is evidence of his faith in its future. He himself must furnish the management and most of the capital until others have enough confidence in his business to be willing to invest in it.

Collateral

Businessmen who have a high credit standing do much of their borrowing on an unsecured basis. Others are often obliged to back up their credit standing with collateral. This is especially likely to be true of a new small-businessman. If he owns a home or other improved real estate, life-insurance policies with a cash surrender value or marketable securities, he may be able to use such assets as collateral for business loans.

Before borrowing on these terms, however, the small-businessman should consider the consequences to himself and his family if he should be forced to withdraw from the business before it becomes firmly established. A small-businessman who retires from business prematurely usually does so at a loss.

Circumstances

Some factors over which the small-businessman has no control may have a bearing on the granting of a bank loan and its repayment. These include:

- Seasonal character of the business.
- Long-run business changes.
- Level of community business activity.
- Competitive position of the firm.
- Nature of the service or product.

Coverage

Proper insurance coverage is extremely important. Small-businessmen are subject to possible business losses from many causes, such as these:

- The death of an owner, partner or principal stockholder.
- Physical damage or interruption of operations as the result of fire, explosion, flood, tornado or other disaster.
- Theft, embezzlement or other acts of dishonesty by owners, officers, employees and others.
- Public liability not covered by workmen's compensation insurance.

A new small-businessman may not be able to insure his company as fully as the owner of an established business, but he should recognize the need. A going concern has little excuse for neglecting to establish and maintain adequate insurance protection against basic risks. (See your insurance broker, agent or a group such as the Associated Oregon Loggers.)

WHAT DOES A BANK LOOK FOR FROM YOUR BUSINESS?

A bank wants to know everything it can about your business and you. If you can, give the banker what he wants to know at the same time you request the loan. It will impress him and it will be a great help to your effort.

Of all the things the bank wants to know, financial information is by far the most important. It has been said that facts are the foundation of credit.

The banker uses as his facts the current ratio of your company—that is, the currents assets divided by the current liabilities. If your current ratio is 1:1, the banker knows that you will probably have difficulty repaying the loan.

Another important ratio the banker looks for is the debt-to-worth ratio—that is, your total debt divided by your net worth. If your company has 2½ times more debt than net worth, you may barely qualify for the loan; anything over that usually will not be looked on favorably.

WHAT SHOULD A LOAN PROPOSAL CONTAIN?

A loan proposal to your banker should usually contain the following things:

- The loan application.
- Your statement of ownership.
- The purpose for borrowing the funds and the exact use you intend for the funds.
- How you intend to repay the loan.
- Comparative and projected financial statements.
- Projected profit and loss and cash flow on a monthly basis for the first year of the loan.
- Projected profit and loss and cash flow on a quarterly basis for the second and subsequent years (usually the first three).
- Comments regarding projected items.
- Market analysis (if applicable and necessary).
- Any other operating figures.
- Comparative and projected operating expenses and overhead.
- Comments regarding operating expenses and overhead.
- Increased operating and projection overhead (if any).
- Monthly projected cost of production—years 1, 2 and 3.
- Comments regarding projected cost of production.
- Personal data on you and your corporate officers.
- Personal financial statement for you (and perhaps your corporate officers) and your partner(s).
- The collateral available for the loan (appraisals if possible).
- Pro forma balance sheet.
- Footnotes on the pro forma balance sheet.
- The business history of your company.

There are even a few more things your banker might want to see:

- Your company income tax statements for the past three years.
- Your personal income tax statements for the past three years.
- Your incorporation papers (if any).
- Current operating leases for facilities and/or equipment.

This may seem like a lot of information to provide for a business loan. It all depends on how much you are going to borrow and how much risk the banker thinks he might be taking. Many times, your success will depend on how much the banker likes your personality and your general business sense.

SUMMARY

You will find detail to be very important when communicating with the various financial institutions. Being a good businessperson is not enough. Today you must be an excellent businessperson with your finger on the pulse of the business at all times.

How you get around in the concrete forest is just as important as how you get around in the real-life forest. If you don't have the stomach for it, then hire someone who does. Do yourself a favor; don't pick a close friend or a relative. After a short time, the business will come between you.

Remember: *progress through preparedness and planning.*

REFERENCES

Elgers, Pieter T., and John J. Clark. 1980. *The Lease/Buy Decision.* New York: Free Press.

Hayes, Rick Stephen. 1977. *Business Loans: A Guide to Money Sources and How to Approach Them Successfully.* Boston: CBI Publishing.

Karrass, Chester L. 1970. *The Negotiating Game.* New York: Crowell.

CHAPTER 3

The Evolution of a Logging Business

Many of the loggers I know personally have been logging all their lives. Their families have been in the logging business for four or five generations. They know what they are doing in the woods, but many times they get lost in the office. As their businesses have evolved over the years, some bad practices and habits have been established. Even though the direction of the business has been forward, progress has been erratic and halting. Bad habits are hard to break, but many times they must be broken in order to save the family business.

THE EVOLUTION PROCESS

As a company grows from miniscule to small and then from small to medium size, certain changes must occur. The logging company that is floundering or in deep trouble is usually the one that has failed to make these necessary changes.

One of the first painful changes comes about when the business outgrows the logger's shirt pocket. Yes, many small loggers run their businesses out of their shirt pockets. That pocket is their file cabinet and all company business goes into it until such time as it is transferred into the seat of the pickup or onto the kitchen table.

Usually the rest of the business, aside from those worn, sweaty papers, is committed to the logger's memory. All his contracts are verbal, he trusts everyone and he thinks everyone trusts him.

So, we find that all his equipment locations and the details for running the various job sides effectively are in his head, along with the contracts and the personnel files. We also find the maintenance program, the equipment replacement schedule and the continued growth plans for the coming years all tucked away in his gray matter.

What would this logger's family do if he were to die or to be seriously injured on the job? Who would know what to do with the business? The business would probably cease to function as a viable cost-effective company. It would die with the owner.

If you are still running your business out of your pocket, please don't be offended by my example. I am trying to make a point about the importance of conducting your business in an orderly and well-organized fashion.

Now is the time to get with the program and make some beneficial changes in your business life. Now is the time to *get control,* to grow if you have the desire and to know exactly where you are financially at any given time.

In running a small company, we can all benefit from the experience of others—and particularly from their mistakes. Why should we have to reinvent the wheel? Almost any problem we encounter in business has already been encountered by someone else before us. Remember, King Solomon said, "There is nothing new under the sun" and nobody questions the wisdom of Solomon.

It's possible to bypass the mistakes of others. I can tell you from my own experience that you must outgrow your shirt pocket or the cab of your pickup. You must expand your horizons. How? Here is the best way:

- Take those important papers out of your shirt pocket and put them in the office where they can do you some good.
- If you don't have an office, then turn the papers over to your wife or your bookkeeper.
- Spend some time in your office each day and make sure that the books are being kept in the way your accountant is always after you to keep them.
- Get all your agreements and contracts in writing (if you haven't already done so) and have your lawyer look them over.
- Spend some time with your key office people, your bookkeeper, your accountant and your attorney. Don't just trust everything to them with no questions. Make them talk to you, and make them speak your language. Remember, they work for you. If you don't like their work, you can fire them just as you would any other incompetent employee.
- Know your business from every angle and viewpoint and share that knowledge with your key people and certain members of your family. Don't leave them in the dark if something should happen to you and your marvelous brain.
- Don't be too easygoing and temper your generosity with good judgment and the spirit of a sound mind.
- Don't spend too much time with the "wheeler dealers." Their purpose in life is to get you to buy something—and it is sometimes something you really cannot afford.
- Don't hire people over the phone. Take time to talk to them eye-to-eye. Find out as much as you can about them and make sure you use the materials outlined in Chapter 9 of this book. It is all for your protection.

In logging, or any other business, you need more than a talent. You need a good business head and the tools to make it all come together. The purpose of this book is to give you those tools.

Logging is the roughest business I have ever had anything to do with. It is affected by more random variables than you can shake a stick at: the ever-changing weather, the whims of government agencies, your personnel, equipment reliability, the mills you log for, the changing economy and sometimes even your own moods. I find it virtually impossible to think of any business so affected by as many things that are beyond one's control.

Don't add to this handicap by ignoring the things you *can* control.

THE STAGES OF EVOLUTION

The evolution of your company will usually take place in three stages. Let's call them *small, fair-sized* and *big,* and define them in the following way, for the U.S. Northwest:

- *Small:* A small logging company is considered to be one having 1 to 25 employees.
- *Fair-sized:* A fair-sized company is considered to be one having from 26 to 100 employees.
- *Big:* A big logging company is one with more than 100 employees.

Of course, other geographic areas of the USA or the world often define small, fair-sized and big differently. These definitions are guidelines only.

Going from one stage of development to the next will almost always create a series of serious organizational problems. During each of the stages your company can be in one of two phases: developmental or decline. Your company will be doing one or the other during each stage of its development.

The Small Stage

Creative activities are essential before your company can even get off the ground. Some of those activities are covered in Chapters 1 and 2. Once you have the company off the ground, new management skills will be needed to take your company to its next stage of growth.

You may decide early on that you do not want to grow beyond the small stage. In the logging business there is nothing wrong or dishonorable about this kind of attitude. It will depend on the amount of money you wish to invest and how much work you wish to get involved in doing. Knowing where to stop in your growth as a logger can many times mean the difference between success and failure.

If you decide that you want to grow to the next stage, you must be willing to make the necessary changes in your company structure and to invest the additional funds necessary to achieve your goals.

The Fair-Sized Stage

Once you have survived the first stage of development through capable business management, you are now ready to tackle the next stage. Now you will usually embark on a period of sustained growth. As your company grows, you must begin to delegate some of your authority, you must institute some formal systems to achieve greater coordination and you must institute a formal planning procedure with a good system of intense review.

The Big Stage

As a logger you may wish to stop your growth long before you reach the big stage. As stated earlier, there is no dishonor in knowing when to stop your growth. If, however, it is your desire to continue to grow, you will find that as you get bigger you are going to reach a red-tape crisis. Your organization may even become flabby because of its rapid and continued growth. You must be prepared to face these problems and to take necessary action to keep them from consuming your company from the inside out. You must now begin to make long-range plans and to lay the groundwork for continued growth at a rate your company can handle. This stage of growth imposes greater risks of decline because of the size of your business.

SUMMARY

Only you can decide which stage of development you wish to attain and then plan your actions to achieve that stage of growth.

Remember: *to fail to plan is to plan to fail.*

REFERENCES

Drucker, Peter F. 1980. *Managing in Turbulent Tmes.* New York: Harper & Row.

Levinson, Harry. 1973. *The Great Jackass Fallacy.* Cambridge, Mass.: Harvard University Press.

Longnecker, Justin G. 1969. *Principles of Management and Organizational Behavior.* Columbus, Ohio: Merrill.

CHAPTER 4

Understanding Depreciation and the Investment Tax Credit

You can turn an understanding of depreciation into additional cash for your growing business. Accelerated methods of depreciation allow you to pay a portion of your income tax later than sooner. You then gain free use of that money for one or more years. The additional first-year depreciation allowance available again to small businesses in 1982 and the investment tax credit can also give your business hundreds of dollars in extra working capital. Finally, to make the best decision regarding replacement of equipment (Chapter 5), you must understand how depreciation is calculated.

DEPRECIATION DEFINED

Since machinery wears out and becomes obsolete, depreciation lets you allocate the cost of that machinery, or any other tangible fixed asset that you use for business purposes, over the life of that asset. One can't take depreciation on intangible assets such as goodwill, land or inventories.

Depreciation in an accounting sense differs from the depreciation you may have experienced in the first year of owning a new car. To an accountant, the depreciation expense is an arbitrary figure that is not linked to the market or economic value of an asset. If, for some reason, the market value of a log loader increases, the depreciation expense is not affected.

Accountants call depreciation a non-cash expense. Although it appears as a deductible expense on the income statement and lowers taxable income, no cash must leave the company to pay for a depreciation expense, as it must to pay a supplier's bill. (However, eventually the depreciable equipment must be replaced.) As a deductible expense, depreciation lowers a company's gross profits and thus shields those profits from the corporate income tax. This shielding effect causes most business people to take as much depreciation as possible in the early years of an asset's useful life.

Taking larger amounts of depreciation in the early years means a larger deductible expense, a smaller profit reported to the government (but not necessarily to shareholders, should there be any) and thus, lower tax payments. In the later years of the asset's useful life, the depreciation expense will be smaller and taxes consequently higher. The net result is to postpone the payment of taxes. This postponement means more working capital. Since a dollar saved now is worth more than a

YEAR	STRAIGHT-LINE CALCULATING METHOD	STRAIGHT-LINE DEPRECIATION EXPENSE	DOUBLE-DECLINING BALANCE CALCULATING METHOD	DOUBLE-DECLINING BALANCE DEPRECIATION EXPENSE	SUM-OF-THE-YEARS'-DIGITS CALCULATING METHOD	SUM-OF-THE-YEARS'-DIGITS DEPRECIATION EXPENSE
1	10% x $12,000	$1200	20% ($12,000)	$2400	10/55 x $12,000	$2182
2	10% x $12,000	$1200	20% ($12,000-$2400=$9600)	$1920	9/55 x $12,000	$1964
3	10% x $12,000	$1200	20% ($9600-$1920=$7680)	$1536	8/55 x $12,000	$1745
4	10% x $12,000	$1200	20% ($7680-$1536=$6144)	$1229	7/55 x $12,000	$1527
5	10% x $12,000	$1200	20% ($6144-$1229=$4915)	$ 983	6/55 x $12,000	$1309
6	10% x $12,000	$1200	20% ($4915-$983=$3932)	$ 786	5/55 x $12,000	$1090
7	10% x $12,000	$1200	1/4 x $3146*	$ 786	4/55 x $12,000	$ 873
8	10% x $12,000	$1200	1/4 x $3146*	$ 786	3/55 x $12,000	$ 655
9	10% x $12,000	$1200	1/4 x $3146*	$ 786	2/55 x $12,000	$ 436
10	10% x $12,000	$1200	1/4 x $3146*	$ 786	1/55 x $12,000	$ 219
TOTAL DEPRECIATION EXPENSE		$12,000		$12,000		$12,000

* Switch to straight-line method, which is allowable at any time.

Figure 8. Comparison of three common ways to calculate depreciation expense on a new piece of equipment with an original cost of $12,000, an estimated useful life of ten years, and no estimated salvage value (under the old tax laws, prior to 1981).

dollar saved later, you should try to defer your taxes even if you have no pressing need for additional working capital.

On August 13, 1981, President Reagan signed into law the Economic Recovery Tax Act of 1981. This legislation, designed to stimulate economic activity, also provided significant and much-needed tax relief for almost all sectors of the economy in the U.S.

The act created a new depreciation method called the Accelerated Cost Recovery System (ACRS), which applies to assets put in service after December 31, 1980, and before January 1, 1985. It also repealed the additional first-year deduction for equipment placed in service in 1981. And, ACRS also affected calculation of the investment credit. However, ACRS applies only to equipment/property acquired after 1980. For earlier purchases (and for certain property specifically excluded from ACRS), the previous rules and the traditional methods of calculating depreciation apply. The following discussion is divided into two sections: pre-1981 rules and ACRS rules, for simplicity.

A REVIEW OF DEPRECIATION RULES: 1980 AND BEFORE

Before 1981, the depreciation allowance was calculated by allocating the cost of an asset (minus its estimated salvage value) over the useful life of the asset. Problems arose in defining *useful life* and *salvage value* as well as in choosing among depreciation methods such as straight-line and double-declining balance.

In calculating depreciation, you had to estimate both useful life and salvage value. In estimating either, you could use your own experience, the experience of your industry or published U.S. government guidelines. Of course, an asset's useful life often differed from its actual physical life. You could not depreciate an asset below its salvage value, but you could generally ignore salvage value unless you anticipated it to be greater than 10% of the asset's cost. Several years ago, the IRS permitted businesses to shorten the estimated useful lives of assets by up to 20%. Ask your accountant for details about the Asset Depreciation Range.

Methods of Depreciation

Straight-line, double-declining balance and *sum-of-the-years'-digits* are the most common methods used in calculating depreciation. Double-declining balance and sum-of-the-years'-digits are both called accelerated methods because they result in a larger depreciation expense in the early years of an asset's life and less in the later years. Whatever the method, the total depreciation expense cannot exceed the asset's cost (less the salvage value for assets purchased before 1981). To use an accelerated method before 1981, the useful life of the asset had to be three years or more, and you normally also had to be the original user of the asset.

To show the difference among the three methods, let's calculate the depreciation expense under each method for a new piece of equipment purchased before 1981 with an original cost of $12,000, an estimated useful life of 10 years and no estimated salvage value. The comparison is shown in Figure 8.

With the *straight-line method,* allocate the $12,000 cost equally at a rate of 10% per year over each of the 10 years of useful life. In each year, the depreciation expense is a simple $1,200.

The *double-declining balance method* is more complicated. First, double the rate calculated for the straight-line method. In our example, this gives a rate of 20%—10% doubled. In the first year, take 20% of $12,000, or $2,400, as the depreciation expense. In the second year, however, take 20% *of what remains* after subtracting the depreciation you took in the first year. The remainder after every subtraction is known as the asset's *book value.* In our example, in the second year you would take 20% of $9,600 ($12,000 − $2,400 = $9,600), which would give you $1,920. The figures for all 10 years and how they are calculated are shown in Figure 8.

If the useful life of the asset were 5 years, the straight-line rate would be 20% per year, and the double-declining balance rate would be twice that, or 40% of the book value each year. Since it is mathematically impossible to depreciate down to zero with the double-declining balance method, the IRS permits you to switch from the double-declining balance to the straight-line method at any time without permission. The reverse, however, is not allowed. Companies generally make the switch to the straight-line rate on the remaining balance when it would give a larger depreciation deduction than the double-declining balance rate. In our example, this point occurs in the seventh year.

Variations on the double-declining balance method include the 150% and 125% declining-balance methods. With either variation, figure what percentage you would take under the straight-line method and then multiply that percentage by either 150% or 125%. In our example, where the straight-line percentage is 10%, the multiplier would be either 15% or 12.5% instead of the 20% we calculated with the double-declining balance.

The *sum-of-the-year's-digits method* is also more complicated than straight-line, but as with double-declining balance, the extra effort to calculate it is generally well worth the trouble. In our example, the useful life is 10 years. The digits for 10 years are 1, 2, 3, 4, 5, 6, 7, 8, 9, 10 and they add up to 55. Fifty-five, then, is the sum-of-the-years'-digits. To figure the depreciation, you reverse the digits; 10, 9, 8, etc., to 1. In year 1, take 10 as the numerator and 55, as you will in each year, as the denominator, giving you 10/55.

Multiply $12,000, the original cost of the equipment, by 10/55 and you get $2,182, which is the depreciation expense for the first year. In year 2, the numerator is 9, and you multiply $12,000 by 9/55, which equals a depreciation expense of $1,964. The multiplier in the third year is 8/55, and so on down to 1/55 in the tenth year.

First-Year Additional Allowance

Prior to 1981, in an attempt to stimulate investment by small businesses, the government permitted them to take an additional 20% of the cost of many assets as a depreciation expense in the first year. There was a maximum of $10,000 worth of assets in a single year, or $20,000 on a joint return. The useful life had to be 6 years or longer and the asset had to be "tangible personal property," which basically excluded real estate. For example, if you decided to purchase a $10,000 truck, you could deduct $2,000 and then figure your normal depreciation on the remaining balance of $8,000 using either straight-line or an accelerated method. You then added that regular depreciation to the $2,000 extra allowed, for a very sizable depreciation expense.

DEPRECIATION UNDER THE ACCELERATED COST RECOVERY SYSTEM

The Economic Recovery Tax Act of 1981 replaced the traditional depreciation system discussed previously, which allocated the cost of an asset over its estimated useful life, with a simplified cost recovery system. This new system permits the recovery of an asset's cost over a shorter period and virtually eliminates conflicts with the IRS over the allowable deduction.

The new system applies to "recovery property," which means *tangible* properties used in a business or held for the production of income and acquired after 1980. Intangible assets and properties acquired before 1981 are still subject to the old depreciation rules. Also, recovery property does not include properties not subject to wear, tear and exhaustion, such as land and corporate securities.

The first step in computing the deduction allowed under the Accelerated Cost Recovery System (ACRS) is the division of recovery property into the following classes:

- *Three-year property.* This class includes:
  - Automobiles.
  - Light trucks.
  - Tangible personal property used in research and development activities.
  - Tangible personal property with a life of four years or less using pre-1981 guidelines for estimated life.
- *Five-year property.* This class includes:
  - Machinery and equipment not included in the 3- or 10-year classes.
  - Single-purpose agricultural structures.
  - Storage facilities used in connection with the distribution of petroleum products.
- *Ten-year property.* This class includes:
  - Depreciable real property with a present class life of 12.5 years or less.
  - Recreational facilities and theme park structures.
  - Manufactured homes.
- *Fifteen-year real property.* This class includes real property other than those items specifically designated as being in the 5- or 10-year classes.

In most businesses, depreciable tangible personal property will be either 3-year or 5-year property and the buildings will be 15-year property.

Once recovery property is assigned to one of the above classes, determination of the ACRS deduction is made by multiplying the unadjusted basis (cost) of the property by the percentage provided in the table below. Note that the percentage to be applied depends on the class to which the asset is assigned and the number of years since the property was placed in service by the taxpayer. The half-year convention is used for all assets in the year of acquisition. Regardless of the date the asset was actually placed in service, the taxpayer will deduct only half a year's depreciation in the first year. Also, salvage value is ignored in the ACRS system. This table, which is in effect for recovery years 1981 through 1984, is based on the 150 percent declining balance method of depreciation. Beginning in 1985, recovery is further accelerated by using the double-declining balance method, as shown in the second and third tables.

If the Recovery Year Is	**The Applicable Percentage for the Class of Property Is**		
	3-year	*5-year*	*10-year*
1	25	15	8
2	38	22	14
3	37	21	12
4		21	10
5		21	10
6			10
7			9
8			9
9			9
10			9

Cars and Light-Duty Trucks

Percentage Applied to Unadjusted Basis

Year of Use	*Asset Put into Use During 1985*	*Asset Put into Use in 1986 or Later*
1	29%	33%
2	47%	45%
3	24%	22%

Most Tangible Personal Property

Percentage Applied to Unadjusted Basis

Year of Use	*Asset Put into Use During 1985*	*Asset Put into Use in 1986 or Later*
1	18%	20%
2	33%	32%
3	25%	24%
4	16%	16%
5	8%	8%

The above table does not include recovery rates for 15-year real property. The recovery rates for 15-year real property (other than low-income housing) are as follows:

Year	Use the column for the month placed in service											
	1	*2*	*3*	*4*	*5*	*6*	*7*	*8*	*9*	*10*	*11*	*12*
1	12%	11%	10%	9%	8%	7%	6%	5%	4%	3%	2%	1%
2	10%	10%	11%	11%	11%	11%	11%	11%	11%	11%	11%	12%
3	9%	9%	9%	9%	10%	10%	10%	10%	10%	10%	10%	10%
4	8%	8%	8%	8%	8%	8%	9%	9%	9%	9%	9%	9%

Note: Complete tables for 15-year real property are in IRS Publication 534.

The new ACRS provides that a taxpayer may elect to compute his ACRS deduction by applying a percentage determined by use of the straight-line method of depreciation, instead of the above accelerated percentages. When the straight-line method is elected, the taxpayer may use either the assigned class life or an extended class life, as specified below:

Class	Extended Class Life
3-year class	Either 5 years or 12 years
5-year class	Either 12 years or 25 years
10-year class	Either 25 years or 35 years
15-year real estate	Either 35 years of 45 years

A taxpayer electing straight-line depreciation percentages for one asset in a designated class must use the same percentage for all assets in that class during the taxable year. The taxpayer may, however, elect to use the accelerated percentages for one class and straight-line percentages for another. The half-year convention also applies to taxpayers using the straight-line percentages. Since the rate of cost recovery has already been selected, it cannot be changed during the life of the property.

Under ACRS, as a general rule no depreciation deduction is allowed for the year in which recovery property is sold or otherwise disposed of. This rule applies also where the taxpayer elects the straight-line percentages. For 15-year real property, however, the percentages are based on the month the property is placed in service (as opposed to the year) and a partial recovery allowance is permitted for such property in the year of disposition.

There are a number of other complicating technical provisions related to ACRS. For example, the taxpayer may elect not to have ACRS apply to property that is depreciated under the unit-of-production method or any other method not expressed in terms of years. The act contains numerous provisions that prevent taxpayers from converting assets held before 1981 into recovery property and thereby obtaining large deductions. For example, ACRS does not apply to property owned or used at any time during 1980 by the taxpayer or a related person. Also, the system does not apply to property acquired as part of certain tax-free transfers. The law also effectively eliminates "component" depreciation on buildings placed in service after 1980 by requiring that components be placed in the same class as the buildings to which they apply.

Please remember: Although the ACRS should simplify computation of the depreciation (cost recovery) deduction and reduce conflicts between you and the IRS, assets placed in service prior to January 1, 1981, will continue to be depreciated under the old rules. The two methods will exist side by side for many years to come.

Additional First-Year Depreciation

The additional first-year allowance for depreciation was not available for property placed in service during 1981. This deduction is substantially increased under the act, but not until 1982.

THE INVESTMENT CREDIT

Major changes in the investment credit rules were necessary to accommodate the new capital recovery provisions. The ACRS recovery period determines the percentage of investment in eligible assets that is qualified for the 10% investment credit. Only 60% of the investment in property in the 3-year class qualifies for the credit, while 100% of the eligible investment in property in the 5-, 10-, and 15-year personal property classes qualifies. One exception that must be noted is that commuter highway vehicles in the 3-year class qualify for the full 10% credit.

A new "at-risk" limitation applies to qualified investments. No credit is allowed for amounts invested in new or used qualifying property placed in service after February 18, 1981, to the extent that the amounts invested are not at risk. In general, amounts not considered to be at risk are those debts for which the taxpayer has no personal liability.

For tax years beginning in 1981, the limitation on the cost of used property that may qualify for the investment credit is increased from $100,000 to $125,000. The act also extends the carryover period for unused investment tax credits on ACRS property to 15 years.

New recapture rules apply to the investment credit taken on ACRS property. If recovery property is disposed of or otherwise ceases to be investment credit property, a percentage of the credit taken must be recaptured. The recapture percentage table is shown below:

If Disposed of	**Recapture Percentage**	
	3-year	*5-year*
Within 1 year	6	10
After 1 year	4	8
After 2 years	2	6
After 3 years	0	4
After 4 years	0	2
After 5 years	0	0

SUMMARY

Always consult your accountant regarding any questions you might have about the information in this chapter. He will show you the best ways of seeking your advantage under old and new tax laws.

Remember: *seek professional help on matters you don't fully understand.*

CHAPTER 5

Evaluating Equipment Acquisitions

As small-business managers, it is not always obvious to us how important employee productivity can be. In logging, most of us look at employee productivity simply as the number of logs a side produces in an hour's time.

Productivity affects both profits and cash flow. Increasing productivity without increasing the cost of doing business automatically increases gross profits. It also improves cash flow because less money is paid out to deliver the product.

Most industries handle increased employment costs by raising prices. This is not always easy to do in the logging business, since there is usually some hungry company waiting to get your business by underbidding you on the job. The other alternative is to produce more logs without increasing labor costs—in other words, put out more logs per employee, or increase productivity.

How is that done? Well, you may be able to get your employees to produce more logs per day for a while with the proper psychological approach, but the improvement seldom lasts very long. A more practical way, sometimes, is to buy a new piece of equipment that will make present operators more productive or that will allow less skilled, less expensive operators to produce the same results.

Keep this productivity factor in mind whenever you evaluate the acquisition of a piece of capital equipment. Look for equipment that:

- Performs operations that are now done by people.
- Makes the entire operation more productive.
- Requires minimum operator training and/or expertise.

As a good businessperson, you know that your company must continue to update equipment to remain competitive in the logging industry. This usually means replacing old, worn-out stuff with new, more productive stuff, simply speaking.

In almost every instance, dozens of questions come up and thousands of dollars ride on your decision. If that puts a lump in your throat, don't feel like the Lone Ranger. There are thousands of businesspeople, large and small, who must make similar decisions each working day. Very few have experts to help them in every area of business and production.

The basic conceptions behind the acquisition of capital assets are quite simple and fairly easy to apply. You don't have to be an economist or an industrial engineer to decide whether that shiny new machine will give you a satisfactory return on your investment.

RETURN ON INVESTMENT (ROI)

First, keep in mind that there is only one good reason for adding a piece of capital equipment to your business . . . and that is to increase the overall profitability of your operation. Don't spend the money unless you will get an adequate return on the dollars invested.

Your return on the invested capital should not only exceed the original cost of the equipment but be great enough to give you an advantage that you could not get by investing the same money in another part of your business.

You can establish a realistic goal for return on investment by adding up three factors:

- The value of the interest the money could earn in a safe investment. Based on mid-1982 returns, a reasonable figure might be something like 9%.
- The value of the predicted annual inflation—perhaps 9 to 11% for 1982.
- The value of the risk you are willing to accept. Conservatively let's say 3%.

When these factors are added (using the smaller numbers), we arrive at 9 + 9 + 3 = 21%. That tells us not to make any expenditure for capital equipment unless it is going to give a net return of at least 21% on the investment. Any return less than that in today's economy is definitely a gamble.

There are always those who buy capital equipment without completely analyzing the return it will bring. And there are some who don't buy, or delay buying, because they look *only* at cost. They fail to relate cost to the increased revenue flow that the new equipment is capable of generating.

Either approach can be a mistake, so let's look at how to analyze the return a piece of equipment should bring. This will give you tangible figures on which to base a buying decision.

JUSTIFICATION METHODS FOR NEW EQUIPMENT

Two rough measures are sometimes used to justify the purchase of a piece of equipment. Although I do not recommend that you use either, both are outlined briefly here since you may be familiar with them.

To explain these two rough methods—and a third, recommended method that follows them—let's use one simple set of assumptions:

- Cost of equipment: $10,000.
- Straight-line depreciation, 8 years: $1,250 per year.
- Investment tax credit, 10%: $1,000 first year.
- Net material/labor savings: $8,250 per year. (This is an assumed figure.)

Before- and After-Tax Cash Payback

The simplest rule-of-thumb method is called *before-tax cash payback*. All you do is set a goal. Let's say you want a 3-year cash payback. Then the $10,000 piece of equipment that you are considering must return at least $5,000 a year in *net sav-*

	YEAR 1		YEAR 2	
NET SAVINGS		$8,250		$8,250
LESS DEPRECIATION EXPENSE		1,250		1,250
ADDITIONAL TAXABLE INCOME		$7,000		$7,000
INCOME TAX AT 35%	$2,450		$2,450	
LESS INVESTMENT TAX CREDIT	1,000		--	
TAX TO BE PAID	$1,450	1,450	$2,450	2,450
ADDITIONAL AFTER-TAX INCOME		$5,550		$4,550
CUMULATIVE AFTER-TAX PAYBACK		$5,550		$10,100

Figure 9. After-tax cash payback calculation

ings before taxes. Since we have assumed a saving of $8,250 per year, you will have your payback in less than 2 years, according to this rule.

The second method requires a little more calculating and gives an estimate of *after-tax cash payback.* This is a more accurate method of justification because any savings in materials and/or labor created by the equipment purchase increases your income, which increases income taxes.

Assuming you still want a 2-year cash payback but after taxes, perform the calculations shown in Figure 9. Now, you can see that the net savings after taxes are considerably less than the before-tax figure, although you will still achieve the full cash payback in 2 years.

Using either of these two simplified methods may cause you to miss opportunities that you otherwise might have taken. So let's examine the method we recommend, which gives a more realistic evaluation of any capital equipment proposal.

Cash Flow Rate of Return

Cash flow rate of return, or CFRR, is based on the fact that a dollar is worth more today than it will be a year from today. This is not because of inflation, but because we can invest today's dollar and have $1.08, $1.10 or more a year from now. The value of next year's dollar is that much less because we have lost a year's return by the time we get it. Inflation—which is included in the 20% goal for return on investment—further reduces the value of the dollar.

The CFRR takes a hard-headed approach to investment. It says, in effect, that if you are going to invest now in equipment that will generate savings over several years, look at the value of those savings in today's dollars, because it is today's dollars that you are investing.

In the CFRR analysis, we are interested in the cash flow generated by the new piece of equipment because cash flow is the real measure of success or failure in business. We want to know the present value of the future cash flow.

The present-value rates for a dollar are based on the percentage of return that we expect to earn. You will find the values tabulated in most books on accounting. We will use our 20% return-on-investment goal and assume that the equipment we are considering will generate savings for at least 5 years, a realistic assumption.

Here are the present values of a dollar at 20% for 5 years:

Year 1: $0.833
Year 2: $0.694
Year 3: $0.579
Year 4: $0.482
Year 5: $0.402

This means, in our case, that a dollar earned 1 year from today is worth only $0.83 today. The same dollar earned 5 years from today is worth only $0.40 today.

The CFRR analysis is not complicated if you take it in easy steps. The first step is to determine how much additional cash flow your new piece of equipment will generate each year after taxes. Figure 10 shows you how to do this, based on the assumptions we made earlier.

As you can see, expected additional cash flow is $6,800 for the first year and $5,800 for each of the 4 following years. To keep it simple, the example uses straight-line depreciation. However, a more rapid form, such as the Accelerated Cost Recovery System for all assets you place in service after December 31, 1980, would considerably increase cash flow by reducing taxable income in the first 4 years of the life of the equipment. This would make the investment in new equipment even more attractive.

	YEAR 1		YEARS 2-5	
NET SAVINGS	$8,250	$8,250	$8,250	$8,250
LESS DEPRECIATION EXPENSE	1,250		1,250	
ADDITIONAL TAXABLE INCOME	$7,000		$7,000	
INCOME TAX AT 35%	2,450		2,450	
LESS INVESTMENT TAX CREDIT	1,000		--	
TAX TO BE PAID	$1,450	1,450	$2,450	2,450
NET ADDITIONAL CASH FLOW		$6,800	EACH YEAR	$5,800

Figure 10. CFRR—step 1: additional after-tax cash flow

YEAR	20% VALUES		NET ADD'L CASH FLOW		PRESENT VALUE
1	0.833	X	$6,800	--	$ 5,664
2	0.694	X	5,800	--	4,025
3	0.579	X	5,800	--	3,358
4	0.482	X	5,800	--	2,796
5	0.402	X	5,800	--	2,332
				TOTAL	$ 18,175

Figure 11. CFRR—step 2: net present value analysis

Now that you have calculated the extra cash flows, how much are they worth in today's dollars? Determine that by applying the present values given earlier, as shown in Figure 11.

When you add up the present values of the projected cash flows, the investment looks good because the future return in today's dollars (total present value) is 80% greater than the cost of the equipment in today's dollars. All that remains is the final step: to determine the cash flow rate of return of this investment, using the method given in Figure 12.

You simply subtract the present value of the investment—the cost of the equipment today—from the present value of the future return. That gives the net present value of the return on the investment. Dividing this by the total present value of the future return produces the CFRR.

PRESENT VALUE OF FUTURE RETURN	$18,175
LESS PRESENT VALUE OF INVESTMENT	(10,000)
NET PRESENT VALUE OF RETURN ON INVESTMENT	$ 8,175
CASH FLOW RATE OF RETURN (CFRR) = C/A = APPROXIMATELY 45%	

Figure 12. CFRR—step 3: calculating the actual return

In this example, the cash flow rate of return is about 45%, or more than twice the 20% goal. This is the final indication that the proposed investment is a good one.

Here is a rule-of-thumb you can use for a quick evaluation of capital equipment proposals. (If a proposal looks good in the basis of this quick assessment, protect yourself by calculating the actual CFRR before accepting it.) This rule says that if you require a 20% return on investment over a 5-year period (using straight-line depreciation and a 10% investment tax credit), every $10,000 in invested capital must show an annual before-tax saving of about $5,500.

THE COST OF SAYING NO

What if you evaluate an acquisition and come up with a favorable CFRR like the one we have just described but still decide not to buy? Technically, you have incurred a cost to your company. Although this "lost opportunity cost" does not appear on the books, it is nevertheless real because it represents money lost by failing to take advantage of an opportunity.

The lost opportunity cost is the difference between the return you could have earned had you purchased the equipment and the return that the same amount of capital would earn in the same period from a conventional investment at 6 to 8 percent. In our CFRR example, $10,000 invested in equipment would earn about $8,000; the same amount invested at 8% would earn only about $4,700. The difference—more than $3,000—is the lost opportunity cost.

BUY THE BARGAIN OR BUY THE BEST?

When faced with a choice between the least expensive and the most expensive equipment, too many loggers look at the price and not at the value (or return) received. It is easier to see the obvious differences in price than to ferret out the real return on investment. If it is profits you want, however, never buy the less expensive of two pieces of equipment until you are really positive that the less expensive buy will also be the less expensive to own and operate.

Consider a simple example. For a certain type of equipment, you get two proposals, one quoting a $15,000 price tag and the other quoting a $20,000 tag. That is a clear difference of $5,000. But what does that mean in the long run? If both pieces of equipment have a life expectancy of 5 years, the cost difference is $1,000 per year. Divide that by a 250-day working year and you are down to a difference of $4 per day.

Now let's assume a hypothetical true labor cost of $5 per hour and say that there is no difference in material or maintenance savings between the two machines. On this basis, you can see that if by chance the $15,000 machine is one hour less productive each day, the $20,000 machine becomes the better buy because its higher productivity more than offsets the cost differential between the two pieces of equipment.

So when comparing two pieces of equipment, look at the real return each will give, not just at the purchase price.

PROS AND CONS OF REPLACEMENT

On the surface, it seems fairly easy to justify buying new equipment that represents a large increase in production or a major savings in employee time. Deciding to replace an old piece of equipment with a newer model is more difficult because the savings are more elusive.

Many companies wait too long to replace a piece of equipment and as a result, lose productivity and profits. There seem to be two reasons for this:

- The belief that a piece of equipment should be fully depreciated before it is replaced.
- The belief that a fully depreciated piece of equipment that is still functioning is saving the company money.

Both of these beliefs are false.

There is no rule that says you must fully depreciate a piece of equipment. If you are using a 5-year depreciation period but the equipment is worn out in 3, replace it and make an adjustment on your income statement. All it means is that the equipment wore out faster than anticipated; you now depreciate it accordingly.

By the same token, a machine that is fully depreciated but still running isn't necessarily saving money, especially if it is becoming more expensive to maintain and repair and isn't as productive as a newer machine would be.

Whether to rebuild or buy new is a tough question and your decision may well be based on the difference in cost. Rebuilding can be like putting a new engine in an old truck—it runs fine, but then you begin to have transmission problems or rear-end problems. If the truck is worn out, it should be replaced, not rebuilt. Compare the value return for purchasing new and you will often find that the cost difference doesn't warrant rebuilding. But if the newer equipment offers no distinct advantages over the rebuilt machine, the option to rebuild may be the better one.

Repair-or-Replace Checklist

You can make up a relatively simple comparison checklist that will help you make the repair-or-replace decision. The checklist should contain the following items. (We are comparing new equipment with old equipment, so make two columns for entering the data next to the items listed.)

- *Depreciation expense* (new equipment cost minus projected used value divided by trade cycle years).
- *Interest expense* (financed amount times interest rate).
- *Fuel cost* (miles times cost per gallon divided by miles per gallon or hours times cost per gallon divided by hours per gallon).
- *Maintenance costs* (cost per mile times miles per year or cost per hour times hours per year).
- *Tire expense* (if equipment has tires). New equipment obviously has new tires; what are you spending on the old?

- *Insurance expense* (yearly cost, new versus old).
- *Personal property tax* (will vary from state to state).
- *Unscheduled downtime* (rule of thumb: 1% times the age of the equipment. Loss-of-production costs per hour).
- *Miscellaneous expenses* (anything not mentioned above).

Now add up all the expenses for new and for old equipment.

Next, look at the tax savings offered by the new equipment:

- *Expense tax savings* (total expense times tax rate).
- *Investment tax credit* (6% of the cost of cars and light-duty trucks and 10% of the cost of most other eligible property in the year of purchase).

Add the two items under tax savings together and subtract this total from the total expenses for the new equipment. That total will be the net expense for the new equipment. Compare that with expenses for the old equipment.

Remember, one of the compelling personal and business reasons for buying a new piece of equipment is the certainty that prices will never be lower than they are right now and that they will in fact continue to escalate. Based on operating and maintenance costs alone, the savings involved in running the older model another year could well be absorbed by an increase in new equipment prices a year from today.

BUY OUTRIGHT, BORROW OR LEASE?

Once you have decided that new capital equipment will provide a satisfactory return on investment, your next question may well be "Should I pay cash for this new equipment, put it on long-term financing or lease it?" Unfortunately, there is no one, simple answer. The choice really depends on the capital and equity structure of your business as well as on cash flow. After some extensive research and conversations with many people, here are some pros and cons concerning the options available to small businesses.

Outright Purchase

Paying cash can be either the best choice or the worst. On the good side, the new equipment purchased immediately becomes part of the owner's equity. There are no finance charges, and you keep all the operating savings realized through the new equipment's performance.

On the other side of the coin, a cash payment means a large, sudden drain on dollars. Unless your company's cash position is sufficient to cover the drain along with all its other cash needs, you could be in for serious cash-flow problems. Remember, too, that the anticipated savings come month by month, not in a lump sum. The future savings will not be worth as much as the dollars being paid out today. Of course, if you have tucked away most or all of the cash purchase price in some form of savings investment at a moderate interest rate, you may be able to get a much better return by withdrawing it for the cash purchase, provided you have carefully considered the return on investment, as previously discussed, in justifying the acquisition. You must be careful when making the decision to pay cash. Weigh all the factors carefully.

Long-Term Debt Financing

An alternate choice might be debt financing. Debt financing has some advantages, which are usually termed favorable financial leverage. You are replacing a single, large cash outlay with a series of time payments. This helps cash flow because the revenue that is freed up by the operating savings will help offset interest on the time payments.

Also, since the interest is tax-deductible, you pay a true interest lower than the stated rate. For example, if you could borrow at 12% and the tax rate is 35%, you pay a true interest cost of 0.12 x 0.65, or 7.8%. (Always negotiate the lowest possible interest rate regardless of this little fact.)

The negative side of debt financing has several aspects. First, there is the total amount of debt carried in relation to net worth. Too high a ratio of debt-to-net-worth is one of the principal reasons many companies fail. Debt reduces the control you have over your company and, if the ratio approaches 1:1, your creditors are in a strong position to take over your company.

Second, the effect that monthly payments will have on cash flow must be considered. Assuming that total debt is not excessive, a sizable purchase can still have a negative effect if you have failed to evaluate and justify the purchase carefully. In other words, if the new equipment does not generate the anticipated savings, there won't be enough cash inflow to offset the outflow of monthly payments.

Last, but not least, try not to use short-term borrowing to cover large equipment purchases. The higher interest rates and larger payments place a heavy strain on your cash flow.

How long to finance depends on the cost, the return on investment and cash flow. The longer you finance, the more you pay in interest. So, once again, careful evaluation of the equipment savings is important.

Another important thing to remember is that whether you pay cash or finance, you usually get title to the equipment on the day of delivery. This gives you the right to claim depreciation and take the investment tax credit on your own tax return. Make sure *you* take these advantages, not the equipment dealer.

Leasing

Leasing almost always costs a little more than buying. But as interest rates continue to remain high, the difference in cost between leasing and installment buying diminishes. Leasing and buying have many things in common. Here are the major points to consider:

- You get the equipment you want, when you want it.
- Your freedom to use the equipment is similar.
- New equipment can be serviced and maintained by the same people furnishing the equipment.

Chevrolet Sedan

BUYING:

Cost for a 36-month installment purchase at 13¼%. For comparative purposes, assume that the automobile is sold after two years.

Downpayment.........................$1,140.00
Monthly payments for 2 years........$5,210.00
Loan still to pay off after 2 years.$2,480.00
Approximate resale value............$4,535.00
Net outlay..........................$4,305.00

LEASING:

Cost for a 24-month closed-end lease. Assume that maintenance and other options are negotiated separately.

Security deposit (refundable).......$ 500.00
Total monthly payments..............$4,885.00
Net outlay..........................$4,885.00

Figure 13. Buying vs. leasing comparison: sedan

- Both share the convenience of spreading monthly payments over a period of years. (See Figure 13 for a comparison.)
- Both methods include interest in each monthly payment.
- Depreciation is incurred whether you lease or buy. Although the accounting for depreciation usually differs for leasing and for buying, it is a real and constant cost in the use of equipment.

Next, let's look at how leasing differs from buying. Again, there are six major points to consider:

- In most cases, the security deposit required to lease is considerably lower than the cash downpayment or trade-in equity needed to finance the purchase of a new piece of equipment.

- With leasing, depreciation is figured into every monthly payment because the lessor anticipates the market value of the leased equipment when it is sold following the expiration of the lease contract.
- Leasing "fixes" the number of months you will use your equipment; on renewal of a lease contract, you can begin to use a brand new piece of equipment. As a result, you can probably avoid some of the parts and maintenance expenses more common with older, heavily used equipment.
- With leasing, insurance, regular maintenance, license plates, sales taxes and other relates expenses can often be prorated in each monthly lease payment over the life of the lease. This will depend, of course, on the individual leasing company's policies and on local regulations.
- If the leased equipment is a vehicle used for business, you can claim income tax deductions. Leasing offers a simple and accurate record of expenses.
- Finally, with leasing, the trade-in process is usually eliminated. You move from equipment to equipment more simply. *Note:* When you first convert from buying to leasing, your equipment dealer can often help you to obtain cash for the equity in your existing equipment. You may wish to channel that cash into other areas of your company.

As you can see, leasing is different from buying. It is an alternative that can offer advantages if you are careful—one that is always worthy of examination and full understanding before you acquire your next piece of capital equipment.

SUMMARY

Now that you have seen the good aspects of each alternative for acquiring new equipment, you should be able to make an accurate decision as to which method you wish to use.

Remember to negotiate, regardless of which way you choose to go. There is extreme flexibility with each dealer. There are some pieces of equipment, however, that cannot be bargained for. If the dealer is the only one who handles that particular piece of equipment, you may be stuck with the price he dictates. But generally, *shop around. It's worth the time.*

REFERENCES

Elgers, Pieter T., and John J. Clark. 1980. *The Lease/Buy Decision.* New York: Free Press.

Hayes, Rick Stephen. 1977. *Business Loans: A Guide to Money Sources and How to Approach Them Successfully.* Boston: CBI Publishing.

Karrass, Chester L. 1970. *The Negotiating Game.* New York: Crowell.

CHAPTER 6

The Business Turnaround

Sometime during the life of your company, you may find yourself in serious trouble. You may have to turn a loser into a winner. You may have to face a turnaround situation. Whether your company recovers, continues to stagnate, or completely disappears from the scene will depend on whether you can effect a turnaround successfully.

A turnaround situation is an abnormal period—a time when things begin to get out of control. You will find that many old and proven management methods no longer work for you or fail to apply to your situation. At this critical stage, doing things by the book will not save your company. You will have to be innovative and receptive to many changes.

All businesses have problems; they vary only in the number and the complexity. To pull off a successful turnaround, you must solve the problems and provide a stable foundation for the renewed and continued growth of your company.

You will find that your options are limited in a turnaround situation. A turnaround involves a number of negative forces, such as a general lack of resources and exceptional time pressures. Turnaround calls for a penetrating viability analysis and a steadfast adherence to your objectives. Unfortunately, there are no standard formulas that can be passed on because a successful turnaround boils down to you, the human equation. The main key is you, the visible hand of management.

BUSINESS FAILURE AREAS

The failure of a business can show up in one or several of four areas:

- Social failure.
- Economic failure.
- Legal failure.
- Managerial failure.

Social Failure

The performance of an organization may be judged by its success in acquiring, developing, and utilizing people. When an organization creates a situation in which its people are not utilized properly and the wrong people are being utilized in areas of extreme importance to the company, this can create a social failure. This type of failure is strictly an internal problem and must be dealt with quickly. Many times, heads must roll to stem the tide of unhappiness and inefficiency that can consume a company from within. Many times the main problem may be you, the boss. You must realize your limitations. If you find that the organization has gone beyond your expertise, get yourself a manager who can take the lead for you and reverse the trends you have caused by your one-man rule.

Economic Failure

Any company can be an economic failure for years and not know it. In the absence of legally enforceable debt, the company might be able to meet its current obligations and avoid legal failure. Economic failure is a situation in which the realized rate of return on invested capital is substantially lower than prevailing rates on similar investments on a continuing basis. This is not an irreversible situation. We will cover several solutions as we progress in this chapter.

Legal Failure

A company is a legal failure when it cannot meet the demands of its creditors and they seek action against it to secure their receivables. As a business owner, you may discontinue operations for a variety of reasons, such as loss of capital, inadequate profits, ill health, or retirement, but if your creditors are paid in full, you will not be tallied as a failure. There are several options open to you even when this situation presses in on you. We will cover them later in this chapter.

Managerial failure

Decline is business failure from a managerial standpoint. Your business can be a failure from this standpoint long before it becomes an economic or a legal failure.

Decline means several years of deteriorating profits. Decline usually includes one or more years of unprofitable operations, large nonoperating write-offs or both. This decline can usually be traced to inadequate leadership in the overall management of the company. Many times it can be traced to one specific area of production, finance or some other vital function within your company. Again, this is not an unsolvable problem in itself.

TYPES OF CORPORATE FAILURE

Basically there are three types of corporate failure:

- *Type 1* companies are small companies that have never risen above a poor level of performance. They usually have a very short life span and they come and go by the hundreds each year.
- *Type 2* companies fly high and grow to fantastic heights only to come crashing down again. These companies are usually founded by a dynamic individual with limited ability. When he reaches his level of incompetence, the company nosedives to oblivion.

- *Type 3* companies are usually mature, professionally managed companies that are often sluggish and out of touch with their markets or the needs of their customers. They usually have a number of management and control defects. Decline usually takes place when several internal and external events occur at the same time. These detrimental events would not create a problem if they happened one at a time; however, they can be devastating when they happen simultaneously.

THE THREE STAGES IN A FAILURE

Regardless of the type of corporate failure, mature companies generally go through three stages on their way to failure:

- The initial collapse
- A plateau
- The final collapse

The length of the plateau stage is directly related to the strength of the balance sheet your company has built up over the years and the timing, size and type of projects that you attempt to get your company off the plateau. The plateau stage is the place to begin the efforts to save your company. Don't wait until you reach the final collapse stage; it may be too late to do anything to save your company.

PREDICTING CRISES

Many times it is possible to predict the coming of crises. Your response to a predictable series of organizational and people crises is often the deciding factor in determining whether your company will die early or go on forever.

Companies that get into trouble are the ones that cannot cope with the internal challenges created by change. This is a failure to recognize the various crises that occur in the organizational life cycle of any company. It can be called functional blindness: you cannot solve the problems because you refuse or fail to see them. The seeds of future financial crises are sown by your failure to cope with people and organizational problems as they occur. In your haste to grow, you may overlook critical organizational issues that must be cared for by management. This can be the beginning of the end.

IS DECLINE INTERNALLY OR EXTERNALLY CAUSED?

In most cases business problems are internally caused. This points out that your responsibility goes beyond protecting your company from the adverse consequences of external challenges. It includes getting your company into a posture, from both production and financial points of view, in which it can resist normal business hazards and other more serious externally caused challenges.

As a logger, you are aware of the sudden changes that can happen in the external environment. These changes take place much more frequently than does sheer bad luck. Most of the time these external challenges are beyond your control in the short run. But remember this: normal business hazards can never be a valid excuse for failure. In a survey conducted by Donald B. Bibeault and reported in his book *Corporate Turnaround,* chief executive officers indicate that external factors do cause decline, but that they are the sole cause only in about 9% of the cases. In another 20% of the cases, decline is caused by both external and internal factors. About 70% of the time, decline is internally caused, although in some cases internal weaknesses are triggered by outside forces.

Some External Reasons for Decline

In today's unstable planning climate, we find that unions can shut down whole industries overnight and government regulations can force you to spend time and money on pollution control, noise abatement, safety and health improvements, slash disposal and a host of other tasks. Consumer organizations can suddenly place artificial pressures on pricing or quality; energy and raw material shortages can interfere with production or create great variations in costs; our Congress can increase business incentives and take them back quicker than you can blink an eye.

As you know, a company does not operate in isolation; it acts on and reacts to a very complex, ever-changing environment. Not to see certain events and calculate their consequences or not to take note of trends and their implications to your business are sure and certain paths to failure.

These external factors can be divided into five categories:

- Economic change.
- Competitive change.
- Government constraints.
- Social change.
- Technological change.

Economic Change. During good times you can get away with many things. A boom covers a multitude of sins, and a bust lays bare many weaknesses. Economic problems can come in many forms: a slackened demand for your product, devaluation of currencies, interest rate hikes, credit squeezes, and even international monetary situations. Right now, inflation and high interest rates are by far our most severe economic problems.

Although economic change can cause decline and failure, you must remember to keep its role in the proper perspective. Good management can offset poor economic conditions.

Competitive Change. Like most businesses, logging operates in a world of constantly shifting competition. If you don't keep your ear close to the ground, try to calculate and quantify the consequences of change and take the appropriate action, then you are certain to lose your competitive edge. You must constantly monitor the outside environment and watch the manipulations of your fellow loggers and the changes they are making. Your competition can cause your decline if you neglect to watch them and anticipate their manipulations. Competition is a positive force; it makes you assume a position of higher production, better equipment utilization and better utilization of your people. You must always look at change as an opportunity, so you can be farsighted enough to recognize the need for changes in your company at the proper time.

Government Constraints. Politicians are playing a growing role in business all over the world, and logging is not immune. Government constraints are relative to your business, whether they are at the local, national or international level. They can affect your raw materials, markets and finance.

The government affects your corporate life through taxation, involvement in your employee records, welfare legislation, pollution control, employee safety and many environmental issues too numerous to mention. Increasing government regulation demands that you use some outside service organizations such as consultants, auditors, lawyers, tax advisers, accountants, advertising agencies and even public relations firms. The internal workings of your corporation must be geared to interact with these outside agencies.

Government is not the main villain in corporate failure. We are told that government causes only about 16% of the externally caused failures, which amounts to less than 4% of the total failures in business.

Social Change. The current movement toward job involvement, job participation and job satisfaction is profound. Failure to understand this and respond to it is to risk a breakdown in employee communication. If you are an astute manager, you can keep up with all these changes and keep your people happy. You should become familiar with new and innovative management techniques that will help motivate your personnel.

Technological Change. Technological progress in the past decade can stagger the imagination. New equipment is being developed every day that allows you to both increase production in your operation and simplify the elements of timber harvesting.

If you are a victim of technological change, you are only a victim of your own lack of foresight. The rate of technological change has not accelerated so fast that a well-managed company can't keep up with it. A poorly managed company, however, will have a very difficult time.

Some Internal Reasons for Decline

Internal factors cause the vast majority of corporate declines. Bad management is the most frequent reason for decline.

Bad management exists because of incompetence, narrow vision and misdirected activity. When you uncover bad management, you will usually find one or more of these reasons in evidence.

Most Common Errors of Bad Managers. Here are some of the most common errors committed by bad managers:

- Failure to keep pace with changes in the marketplace.
- A lack of sufficient operating controls.
- A lack of control information.
- Ignoring controls altogether (even where they do exist).
- Overexpansion.
- Overdiversification.

- Overlooking business fundamentals.
- Excessive leverage.

WATCHING FOR THE EARLY-WARNING SIGNALS

It is difficult, if not impossible, to discern and diagnose company problems before they threaten the progress or the existence of your company. Business warnings don't make themselves known; they have to be sought out and nailed down. That can be a really tough job for a man who is very close to the production portion of his business and not close to the financial part. Most of the loggers I have known have had this problem of one-sidedness.

When problems are first uncovered, they are usually not clearly defined enough to act upon. Early signals are generally weak signals, and weak signals are symptoms rather than causes. In the last few months before a company collapses, the number and severity of the symptoms rapidly increase. Sometimes a stagnating company needs to be shocked into some constructive action. We are told that eight out of ten early warning signals are ignored by management.

YOUR MOMENT OF TRUTH

The moment of truth comes when you are forced by the crises to realize that things are falling apart. Some companies can rise to the occasion while others just stumble on to reorganization and eventual bankruptcy. You cannot blame uncontrollable external factors for all your woes; this cannot be accepted as an excuse. You must accept the majority of the blame for not admitting your errors and for not watching every aspect of your business.

Maybe you have taken the ostrich approach—putting your head in the sand hoping that everything will eventually work out and it will all return to normal. This is a great way to lose your business and everything you have worked so hard to achieve.

It is hard to ignore your problems when you run out of money. If this happens, you will find that the financial institutions are usually willing to support you, the first time, as a practical matter. The second time they will balk and hesitate, and the third time they will turn their backs and let you drown.

REALITY COMES HOME

Eventually, a company that is taking the big slide downhill has to admit to the seriousness of the situation. This is the time when leaders begin seeking answers rather than excuses and the rats begin to abandon ship. At this point, what you do and how you do it will determine whether your company will survive or not.

The process of pulling yourself up by your bootstraps is a painful one, but it doesn't have to be as agonizing as some make it out to be. Being calm in a crisis is much easier to preach than to practice, but now is the time to be calm. Achieving a successful turnaround is mostly a victory of your management performance. It is also a testament to the courage of your employees, and the loyalty of your customers and creditors.

THE BASICS OF A TURNAROUND

Turnaround means a substantial and sustained positive change in the performance of your business. Change is a hallmark in the turnaround situation. Acceptance of change is critical.

There is a general sequence of events that a company must go through in pulling out of a decline and finally returning to healthy growth. The discrete stages of a turnaround are the following:

- The management change stage.
- The evaluation stage.
- The emergency stage.
- The stabilization stage.
- The return-to-normal stage.

The length of time for each stage varies greatly. It can take anywhere from two weeks (for a small company) to six months (for a large organization) to make value judgments about your business.

Four key elements apply to any business situation. They are:

- A competent management with full authority to make all the necessary changes.
- An economically and competitively viable core operation to hold the whole thing together.
- "Bridge" capital from external and internal sources to finance the business turnaround.
- A positive attitude and motivated people to sustain the momentum of the initial turnaround.

Remember, at this point in your corporate life span "cash is king." Beyond this point, the key to sustained success is positive thinking.

The Management Change Stage

If you are to remain the top manager, you must be willing to make some drastic changes in management processes. You, or someone you choose, must begin to make dramatic changes quickly. You can begin these changes by using a hands-on management style, by delegating absolute authority to your assigned management, by the introduction of tight controls on every aspect of your business and by putting emphasis on good "people motivators."

During the early stages of the turnaround, tight controls over purchasing and head count are critically important. You must control the cash flow pipeline.

With sales and profits down, you must take immediate action to improve the near-term outlook and to develop a base for the long term. An action plan like this has three principal purposes:

- To give evidence of action and change and to make clear to your organization before much time goes by that you are in charge.

- To make short-term improvements—either by cutting your costs or increasing your volume.
- To lay a foundation for both an organized approach to the longer term problems of the company and a control system to tighten the company up.

The key to success lies not in complicated systems of management control but in swift action that accomplishes a few simple things extremely well.

As a successful turnaround leader, you must attack problems, obtain total financial control, find out exactly what is going on in your company, locate your key managers, seek profit improvements, gather all the facts and make the proper decisions. Once you have taken charge, you must keep charge.

Importance of Motivation. Motivation is necessary to sustain a successful turnaround. To motivate people in a turnaround you must first lead them, then kick them in the pants, reestablish a climate of success, get them to participate and communicate well to all during the various stages of the turnaround. Accomplish first—motivate second.

The Evaluation Stage

The evaluation stage of a turnaround entails making a viability analysis of the business and preparing an action plan to solve the problems of the company in the short run.

One of the first things you must do is to order the preparation of the reports and data you will need for the evaluation. This evaluation will go beyond feel and will utilize concrete information and current reports concerning the condition of your company.

Remember the three key elements in accomplishing your turnaround. They are:

- A willingness on your part to face up to the consequences of your troubled situation and to take drastic action.
- A viable core operation that will provide the internal financing required to sustain your company through its turnaround stages.
- A willingness on the part of your lenders to provide the bridge financing needed in the initial stages of the turnaround while your company turns its dormant or poorly performing asssets into the cash you so desperately need.

Evaluating Your Company's Financial Strength. There are four key-issue areas of financial analysis in the evaluation stage. They are:

- The state of your company's balance sheet and the ability of your company to finance the turnaround from internal sources.
- The difference between operating-statement profits and cash flow.
- The need for you to understand completely how your money is made and how it is disbursed.
- The state of your company's financial information system.

Don't spend too much time on the evaluation stage. You as the owner should have a good feel for everything in fairly short order.

The Emergency Stage

In this stage of the turnaround you do whatever is necessary to ensure the survival of your company. Here are a few things to consider:

- *Asset deployment.* Your assets should be broken down into borderline divestment units and main divestment considerations.
- *Financial management.* The main considerations during the emergency stage of the turnaround are:
    Cash flow analysis and control.
    Debt restructuring.
    Working-capital improvements.
    Profitability analysis support.
    Cost reduction support.
    Elimination of creative accounting.
- *Operations management.* Truly significant results can be achieved in cutting operational costs when the pressure is on. The following areas should be considered:
    Shutting down some of your operations.
    Reducing manpower, both direct-labor and overhead.
    Reducing inventory investments.
    Controlling purchasing.
    Increasing your productivity.

The Stabilization Stage

Stabilization takes place after the company has settled down from the trauma of the surgery during the emergency plan. The objectives of your company during the stabilization process are achieved by adhering to three important strategies:

- Concentration on profit improvement by continued attention to margin achievement.
- Enhancement of profits by running the existing operations more efficiently.
- Reposturing your company for increased profit and return on investment on the assets employed in making that profit.

During this stage, you should be striving for liquidity improvement and balance sheet cleanup.

The Return-to-Normal Stage

Your desire to return to normal growth is deep rooted and has helped you to sustain yourself during the painful process of turnaround.

At this stage, your primary objective should be longer range development—positioning your company for the future. This development objective will probably include the following strategies:

- The moving of assets from slow-growth areas to faster growth areas in a way consistent with high margin and return-on-investment standards, thereby closing your strategic performance gaps.

- The positioning of the company's financial capabilities to support increased and continued growth.
- Investment in internal operational capabilities to provide increased productivity and capacity.
- An increased emphasis on expansion, improved visibility and new projects.
- A renewed emphasis on developing your human resources by establishing motivation and development programs for your people.

By the time you reach this stage of your turnaround, the risk has declined and you should begin to place emphasis on revenue growth and margin discipline. Your battle should be over.

Now the difficult task of maintaining consistent growth and profitability over the coming years has begun. May good fortune follow you always.

BAILOUT STRATEGIES SHORT OF CHAPTER 11

When a company becomes realistic after an unsuccessful attempt at turnaround, it has several alternatives short of filing for protection under the Bankruptcy Act. The alternative it chooses depends a great deal on its financial condition and the rate of its operating losses. A company could:

- Cease operation and have an orderly liquidation of assets.
- Sell out as a going concern to an interested buyer who might be able to save the company through greater expertise and utilization of the company assets.
- Seek a merger with a stronger company in the logging business.

THE TOILS OF CHAPTER 11

When an effort to effect a turnaround has failed, a company in perilous condition files under the bankruptcy section of the reorganization section of Chapter 11 of the Bankruptcy Act. Filing under the bankruptcy section is tantamount to an orderly liquidation because it offers little hope that the company will ever resume normal operations as a going concern. Chapter 11 is a last-ditch effort. If you need protection from your creditors, it is sometimes the only way to get it. Chapter 11 can be a nightmare. In most reorganization situations, the creditors are so deeply involved that they become, in effect, owners. But it should be remembered that they will always have different objectives than those of the owner. Your creditors will often want you to sell the winning portions of your viable-core operation in order to satisfy their claims. As you can well imagine, this action also has the effect of guaranteeing the failure of the business.

LOGGER'S POSITIVE ACTION PLAN

If your business has led you into a dark corner, there are several things you can do to get yourself back into the light. Here are some ideas. I like to call them my Positive Action Plan.

- Don't panic. Business has to be guided by logic and positive thinking.
- Curtail all unnecessary spending in all areas of your company (including yourself and your wife).
- Lay off all personnel who are not essential to the effective and safe operation of your company.
- Return any lease equipment not vital to your effective operation.
- Sell anything that can be spared in order to gain as much operating capital as possible.
- Discontinue all overtime (unless you must have it due to the layoff of extra people).
- Establish a cost control system like the one in this book.
- Continually monitor the performance of each individual and each job side to ensure optimum use of men and material.
- Explore all sources of additional operating capital using the ideas put forth in this book.
- Renegotiate your logging prices (if you can). Talk to your customers; many times they will be willing to help in many ways. They will even be more cooperative if you can show them how your business is doing on paper. *Do your homework.*
- Add up the total damages to be taken care of and make some strong, positive moves to correct each of them. Talk to all your creditors and work with them.
- Implement all applicable actions immediately.
- Follow up on all actions taken as a result of this plan.
- Reevaluate the results and continue making positive steps forward based on your increased knowledge of your financial position.

Some of the 14 suggestions may not apply to your business, but in general one or more of these items is the cause of your present problems. Even if your business is not in trouble, it is good to review these suggestions periodically and apply any that would help to make your position stronger.

SUMMARY

This chapter was written for those who are experiencing difficulty in keeping their companies solvent. It is my hope that you will never have to use the information, but if you do, follow the guidelines closely. They have been proven effective for many companies in serious trouble.

Remember: *power comes through knowledge.*

REFERENCES

Drucker, Peter F. 1980. *Managing in Turbulent Times.* New York: Harper & Row.

Bibeault, Donald B. 1981. *Corporate Turnaround.* New York: McGraw-Hill Book Company.

Levinson, Harry. 1973. *The Great Jackass Fallacy.* Cambridge, Mass.: Harvard University Press.

Longnecker, Justin G. 1969. *Principles of Management and Organizational Behavior.* Columbus, Ohio: Merrill.

SECTION TWO
Operations Management and Controls

CHAPTER 7

Simple Aids in Cost Accounting

Most companies have two basic accounting systems. One system summarizes the overall financial state of the company. The other system, the managerial or cost accounting system, measures in detail all of the particulars of people and equipment and puts charges into specific areas and categories.

In this chapter we will deal with the simple aspects of bookkeeping and how they interact with the basic cost accounting system.

THE IMPORTANCE OF ACCURATE RECORDS

Keeping accurate cost accounting records can be simple. The small investment in time will provide you with instant knowledge about the success or failure of your business.

As mentioned earlier, bad record keeping (or no record keeping) is a primary reason for small-business failure. Recent studies of small-business bankruptcies show that nearly one-third kept no records and less than one-fourth kept adequate records.

Most of the failed businesses had no idea they were failing, even though they had been on the road to ruin for months.

YOUR BOOKKEEPING SYSTEM

I highly recommend that you have a certified public accountant (CPA) set up your entire bookkeeping system. Once he develops your books, keeping them current and accurate will be your responsibility.

Your CPA will design a system to give you far more than just good bookkeeping. He knows you will want a continual analysis of your business that will help you with intelligent planning and control.

He will design the bookkeeping system to do the following things:

- Give the results of all operations clearly, quickly and with full explanations.
- Permit quick comparison of current and past costs.
- Quickly "red-flag" all variations from the patterns essential for profits.
- Show possible employee frauds, waste, errors and other irregularities.
- Make sure your equipment is producing properly and not overly expensive to operate.

- Aid in the prompt filing of accurate reports and tax information.
- Help you choose proper production and bidding policy. This will allow you to concentrate on more profitable sales and avoid borderline sales.

The system should be modern, economical and completely functional.

Make sure you have daily, weekly, monthly and annual information when it is needed. Be sure the system gives information swiftly and accurately. Good records and good logging business go together.

THE INFORMATION-GATHERING SYSTEM

Your accountant will provide tools to make your job easier and to allow you to properly categorize all input into your cost accounting system.

To help you separate your expenses, your accountant will prepare a chart or manual of account code numbers for use by your bookkeeper. Many of the account codes will deal with income and expenses. We will focus our attention on these items. What we want to know is this: did we make money this week or did we lose money?

The account code numbers should be written on your check stubs as the checks are issued. This will help you in making journal entries and in properly fixing each cost.

A PARTIAL CHART OF ACCOUNTS

A partial chart of accounts might appear as shown in Figure 14. You will notice that some numbers were left unassigned. This allows you to add something you might have overlooked at a later time.

SOME SIMPLE BOOKKEEPING PRACTICES TO REMEMBER

Earlier we mentioned journals and journal entries. We will not go into all the details of bookkeeping. But we will cover some basic things about making entries in the cash receipts and cash disbursements journals. Your bookkeeper should keep these things firmly in mind in the everyday conduct of the business.

The Cash Receipts Journal

One of the journals your accountant will prepare for you is a cash receipts journal. Your bookkeeper will be making the daily entries, not your accountant.

This journal gives you a record of all the money you have received during the course of your business days, months and year. Each time money is received, an entry is made in this journal. As a result, this journal will tell you when the money came in, where it came from and why it came in.

The entries should be made on a timely basis. The best time to enter the information is when you make out the bank deposit slip.

Code Number	Account Name
401	Logging revenue
402	Equipment rental
403	Employee receivables
404	Miscellaneous income

(We see here that 400 series numbers are usually used for income entries.)

Code Number	Account Name
501	Timber purchases (paid on volume)
502	Logging subcontractors
503	Logging equipment rental
504	Logging equipment maintenance and repair
505	Logging supplies—wire rope and rigging
506	Logging saw maintenance, repair and purchase
507	Scaling fees
508	Roads, fire trails and cleanup
509	Miscellaneous logging expense
510	Equipment relocation
511	Equipment lease
512	Equipment purchase
513	Cutting subcontractors
514-520	(Unassigned)
521	Trucking subcontractors
522	Trucking equipment rental
523	Truck maintenance and repair
524	Licenses, permits and PUC plates
525	Fines and citations

Code Number	Account Name
526	Miscellaneous trucking
527	Tire expense
528	Petro-products (fuel, oil, etc.)
529	Radio and communications
530-535	(Unassigned)
536	Cleaning, building maintenance and sanitation service
537	Utilities
538	Shop supplies
539	Office supplies
540	Dues and subscriptions
541	Postage and shipping
542	Charitable contributions
543	Training and certification
544	Entertainment and travel
545	Advertising and public relations
546	Miscellaneous office expense
547	Legal and professional fees
548	Payroll—officers
549	Payroll—hourly
550	Payroll taxes
551	Personal property taxes
552	Real property taxes
553	Income taxes
554	Other taxes (specify)
555	Insurance (all)
556	Interest paid
557	Reporting penalties and fines
558	Bank charges

Figure 14. Partial list of account code numbers

Notice in Figure 15 how the deposit slip can be made out to help in this regard. The following is a step-by-step explanation of what has been put on the deposit slip.

Under *currency,* one of the employees paid $55.00 cash for some gas taken from the company pump during the week for use in his own car.

Under *checks,* Item 1 is a check from Amazon Wood Products (bank code 96-517) for $257.50. Item 2 is a check from Bangle Veneer Company (bank code 24-12) for $30,400.00. Item 3 is a check from Evergreen Plywood Company (bank code 22-14) for $5,840.00. Item 4 is a check from an employee who has a loan from the company (bank code 96-512) for $120.00. Item 5 is a check from a company renting one of our loaders (bank code 24-42) for $1,200.00.

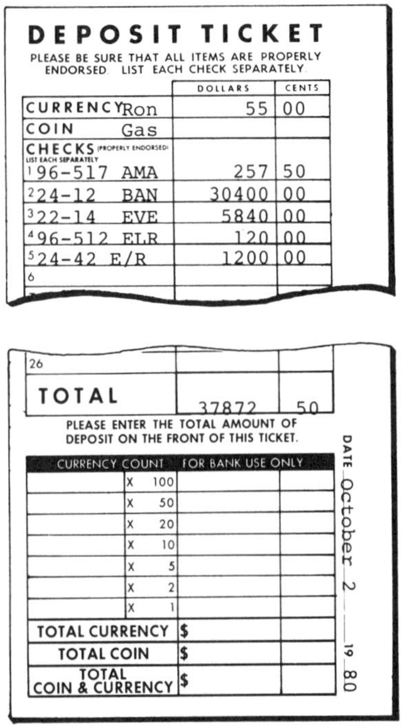

DEPOSIT TICKET

PLEASE BE SURE THAT ALL ITEMS ARE PROPERLY ENDORSED. LIST EACH CHECK SEPARATELY.

	DOLLARS	CENTS
CURRENCY Ron	55	00
COIN Gas		
CHECKS (PROPERLY ENDORSED) LIST EACH SEPARATELY		
1 96-517 AMA	257	50
2 24-12 BAN	30400	00
3 22-14 EVE	5840	00
4 96-512 ELR	120	00
5 24-42 E/R	1200	00
6		
26		
TOTAL	37872	50

PLEASE ENTER THE TOTAL AMOUNT OF DEPOSIT ON THE FRONT OF THIS TICKET.

DATE October 2 19 80

CURRENCY COUNT		FOR BANK USE ONLY	
	X 100		
	X 50		
	X 20		
	X 10		
	X 5		
	X 2		
	X 1		
TOTAL CURRENCY		$	
TOTAL COIN		$	
TOTAL COIN & CURRENCY		$	

Figure 15. Sample deposit slip

Under *total,* we see that all the revenue added up to a total deposit of $37,872.50.

Journal entries. Now notice how the journal entries are made for the same items appearing on the deposit slip. In Figure 16 note that the columns of the ledger sheet are marked so that the left side shows where the money came from and the right side shows what the money was received for.

Now let's begin to make the journal entries from the deposit slip into the cash receipts journal. Here is a step-by-step rundown:

1. The date is October 2, 1980, for each item.
2. The first source of revenue was $55.00 cash from an employee for gas used in his own car. So we put "Ron Macklin" as the source of revenue.
3. The amount received (and deposited) was $55.00.
4. Now move over to column 8 (Miscellaneous Income) and log $55.00 again.
5. Then in column 9 and 10 write "for gasoline."

Your next entry will be made in a like manner:

1. Again the date is October 2, 1980.
2. The second source of income was "Amazon Wood Products."

CASH RECEIPTS JOURNAL For the year 19 80

	DATE		SOURCE OF REVENUE	CK	1 AMOUNT	2 AMAZON WOOD PRODUCTS	3 BANGLE VENEER COMPANY	4 EVERGRN. PLYWOOD COMPANY
1	10	2	RON MACKLIN	✓	55 00			
2	10	2	AMAZON WOOD PRODUCTS	✓	257 50	257 50		
3	10	2	BANGLE VENEER	✓	30400 00		30400 00	
4	10	2	EVERGREEN PLYWOOD	✓	5840 00			5840 00
5	10	2	LOAN PAYMENT (EMPLOYEE)	✓	120 00			
6	10	2	EQUIPMENT RENTAL (NORTON)	✓	1200 00			
7	10	9	AMAZON WOOD PRODUCTS	✓	4250 00	4250 00		
8	10	9	BANGLE VENEER	✓	21000 00		21000 00	
9	10	9	EVERGREEN PLYWOOD	✓	7350 00			7350 00
10	10	9	JERRY BLACK	✓	32 50			
11	10	16	AMAZON WOOD PRODUCTS	✓	7310 00	7310 00		
12	10	23	AMAZON WOOD PRODUCTS	✓	4780 00	4780 00		
13	10	23	BANGLE VENEER	✓	12000 00		12000 00	
14	10	23	AETNA LIFE & CAS.	✓	347 50			
15	10	23	EVERGREEN PLYWOOD	✓	2730 00			2730 00
16	10	30	AMAZON WOOD PRODUCTS	✓	373 00	373 00		
17	10	30	BANGLE VENEER	✓	12140 00		12140 00	
18	10	30	JUNK YARD DOGS	✓	120 00			
19	10	30	EDDIE BIROSONG	✓	12 50			
20								
21								
22								
23			MONTHLY TOTALS		110318 00	16970 50	75540 00	15920 00
24								
25								

Figure 16. Sample ledger sheet—cash receipts journal

3. The amount received (and deposited) was $257.50.
4. The amount received will also be entered under "Amazon Wood Products." This will help you at the end of each month to determine easily and quickly how much you received from each of the mills individually.
5. Next you will make an entry under "Logging Revenue."

Your next entry will be made as follows:

1. Again, the date is October 2, 1980.
2. The third source of income is "Bangle Veneer Company."
3. The amount received (and deposited) was $30,400.00.
4. The amount received will also be entered under "Bangle Veneer Company," again, to help you in your monthly mill totals.
5. Next you will make an entry under "Logging Revenue."

Your next entry will be made exactly like the last two.

TRIANGLE M LOGGING COMPANY — MONTH OF OCTOBER

5	6	7	8	9	10	11	12
LOGGING REVENUE 401	EQUIP. RENTAL 402	EMPLOYEE RECEIVE. 403	MISC. INCOME 404	EXPLANATION			
			55.00	FOR GASOLINE			
257.50							
30400.00							
5840.00							
		120.00		FROM RON MACKLIN			
	1200.00			FROM NORTON IND.			
4250.00							
21000.00							
7350.00							
			32.50	FOR GASOLINE			
7310.00							
4780.00							
12000.00							
			347.50	SETTLEMENT INSURANCE CLAIM			
2730.00							
373.00							
12140.00							
			120.00	REFUND ON BUMPER TRADE			
			12.50	FOR GASOLINE			
108430.50	1200.00	120.00	567.50				

Figure 16. *Continued*

1. The same date.
2. The source is "Evergreen Plywood Company."
3. The amount is $5,840.00.
4. The entry under "Evergreen Plywood Company."
5. The entry under "Logging Revenue."

Your next entry (No. 5) is slightly different because the source of the revenue is different:

1. The same date.
2. The source is "An employee loan repayment."
3. The amount is $120.00.
4. The next entry is made under "Employee Receivables."

The last entry made from this deposit slip is as follows:

1. The same date.
2. The source is "Equipment Rentals."
3. The amount is $1,200.00.
4. The next entry is made under "Equipment Rental."

The total of all entries would be the exact amount deposited on this deposit slip—i.e., $37,872.50. Each time you make a bank deposit, the procedure of making the appropriate entries in your cash receipts journal should be followed. In this way you will have all your revenues accounted for and all your journal entries made in a timely manner. At the end of each month, the columns would be totaled. You would then have the total number of dollars received and the totals from each of the mills you log for each month.

Then on the right side of the page you would have the total Logging Revenue (this total should equal the totals of columns 2, 3 and 4—in this case

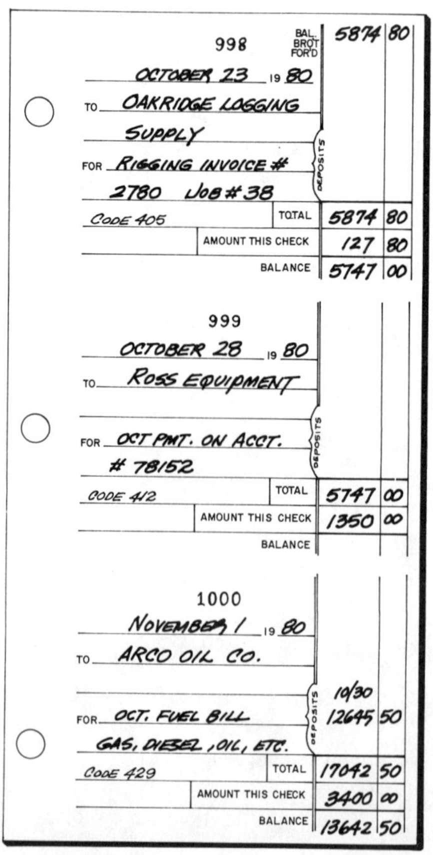

Figure 17. Sample check stub

$108,430.50), Equipment Rental, Employee Receivables and Miscellaneous Income. The totals for columns 5, 6, 7 and 8 should always equal the total revenue received for the month in column 1—in this case, $110,318.00.

After totaling all the columns and balancing out the items as mentioned, you would want to make sure you checked off each entry in column 1 against your actual deposit slips. Use the small column between Source and Amount to put your check mark. After completing the ledger sheet for each month, start a new ledger sheet for the next month. Never put two months on the same ledger sheet; it only creates confusion.

The Cash Disbursements Journal

A cash disbursements journal is necessary to show where the money went. This journal is kept in conjunction with your company checkbook. Each time you write a check, you must make a journal entry. You can do this on a daily, weekly or monthly basis, as you prefer. But, make sure you do it! Keeping your books up to date is an important factor in good management.

It is important to make your entries as detailed as possible, not only in your checkbook but also in your journal. Figure 17 shows how the entries should be made in your checkbook. Where possible, put the invoice number and the job number; also put account numbers and code numbers from your accounts journal. It also helps to put a brief description of what the item was—i.e., rigging, monthly payment, fuel, and so forth. The more detail you put on the check stubs, the easier time you will have making your journal entries.

You will notice in Figure 18 that you should also put some detail on your checks. Later, when you match the checks to your journal, not only will you be able to use the check number, but the additional information will also be helpful.

Making Disbursement Journal Entries. To make journal entries you will need

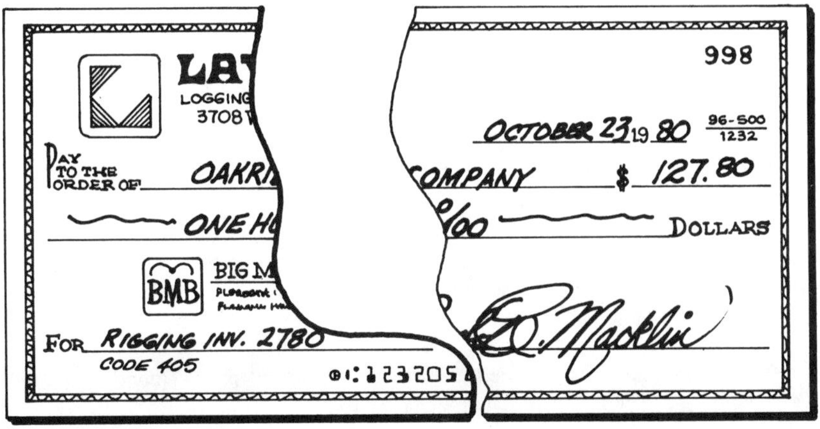

Figure 18. Sample check

CASH DISBURSEMENTS JOURNAL For the year 19 80

DATE		PAYEE	CK #	F	1 CHECK AMOUNT	2 WIRE ROPE 505	3 ROADS FIRETRAIL 508	4 EQUIP. PURCHASE 512
—	—	PREVIOUS PAGE TOTALS			94880 00	4850 00	3600 00	5050 00
10	15	RED RIVER CUTTERS, INC	974		1350 00			
10	15	BOB BIRD & SONS TRUCKING	975		850 00			
10	15	M & M ENVIRONMENTAL	976		1200 00		1200 00	
10	15	ROGER'S MOTORS INC.	977		585 00			
10	16	FLAPMAN BROS. OFFICE	978		125 80			
10	16	'MA' BELL	979		114 00			
10	17	ELECTRICAL COOP	980		78 50			
10	17	LARRY LAMPKIN	981		385 50			
"	"	GREG BROWNELL	982		427 80			
"	"	JOHN BAXTER	983		397 20			
"	"	DAVE HARRIS	984		573 40			
"	"	VOID	985		———			
"	"	ROGER DAVIS	986		750 00			
"	"	RON MACKLIN	987		850 00			
"	"	CARMEN MACKLIN	988		350 00			
10	20	OR. MOTOR VEH. DEPT.	989		20 00			
10	20	ALEX'S TIRE SERVICE	990		450 00			
10	20	TEXACO OIL CO.	991		43 20			
10	20	INDIVIDUAL'S BANK (IRS)	992		1250 00			
10	21	POSTMASTER	993		90 00			
10	21	OXY WELDING SUPPLY	994		57 30			
10	21	COMM. RADIO CO.	995		157 80			
10	22	JEFF'S SAW SHOP	996		52 50			
10	22	PETTY CASH	997		100 00			
10	23	OAKRIDGE LOGGING SUPPLY	998		127 80	127 80		
10	28	ROSS EQUIP. CO. #78152	999		1350 00			1350 00
		(TOTAL THIS PAGE 2-27)			11735 80			
		MONTHLY TOTALS			106615 80	4977 80	4800 00	6400 00
		BANK CHARGES			3 58			
		GRAND TOTALS			106619 38			

Figure 19. Sample ledger sheet—cash disbursements journal

your checkbook and list of accounts. The checkbook will tell you what to enter and the list of accounts will tell you what code to charge.

Looking at Figure 19, you see that the journal is set up very much like the cash receipts journal, only this time we want to know where our money has gone, not where it came from. Figure 19 represents a page taken from the journal at the end of a month (October). There are several reasons for doing this:

- To show you how to carry the totals from the previous pages (months).
- To show you how to log your bank charges.
- To show you how to total out the month's activity and balance the journal pages for that month.

TRIANGLE M LOGGING CO. — MONTH OF OCTOBER

5 CONTRACT CUTTING 513	6 TRUCKING GYPO'S 521	7 TRUCKING MAINT. 523	8 LICENSES PERMITS 524	9 TIRES REPAIRS 527	10 GAS DIESEL 528	11 OTHER CODES	12 AMOUNT	Line
15000.00	17000.00	3000.00	300.00	1500.00	4580.00	—	40000.00	1
1350.00								2
	850.00							3
								4
		585.00						5
						539	125.80	6
						537	114.00	7
						537	78.50	8
						549	365.50	9
						549	427.80	10
						549	397.20	11
						549	573.40	12
								13
						549	750.00	14
						548	850.00	15
						548	350.00	16
			20.00					17
				450.00				18
					43.20			19
						550	1250.00	20
						541	30.00	21
						538	57.30	22
						529	157.80	23
						506	52.50	24
						546	100.00	25
								26
								27
								28
16350.00	17850.00	3585.00	320.00	1950.00	4623.20	—	45759.80	29
								30
								31
								32
								33

Figure 19. *Continued*

The entries are made in the following manner:

- Line 1 shows the previous page totals. During each month, you will usually have several pages of entries. Wherever your last entry for the month falls, total out your page at that place. Never put two months on the same page; it only creates confusion.
- Line 2 shows the *date* of the entry, October 15th. The *payee* is Red River Cutters Inc. The *check number* is 974. The *check amount* is $1,350.00.
- Since the check is written to pay a contract cutter, we look at our manual of accounts and it tells us the code number is 413 (Contract Cutting). So in the Contract Cutting column we make the entry of $1,350.00.

- Each remaining item is logged in exactly the same manner: the date, the payee, the check number and the check amount in the appropriate code number column.

On line 28 you will notice that we put the total for this page only. That can be done on each continuing page if you wish to do so. We are primarily interested in only the monthly totals for each column at the month's end, not page totals.

On line 29 you see the monthly totals for each of the columns. The total for column 1 should equal the totals of all the other columns put together. If these totals match, you have made all your entries in the correct manner.

On line 30 you will see that the bank charges have been added to the month's expenses. Any overdraft charges, charges for checks and service charges should be shown here.

On line 31 you see the total expenses for the entire month of October in column 1. On line 29 the remaining columns show you the monthly totals for all the coded accounts.

Column 11 lists the other codes in your manual of accounts that were not listed in columns 1 through 10. You have room only for the major expenses in the first 10 columns. The smaller accounts are listed in column 11 and the amount of the check is listed in column 12. This keeps you from having a ledger sheet 4 feet long. Once you have the sheet completed, you begin a new sheet for the next month.

The Petty Cash Journal

Looking at Figure 19, line 25, you will notice an entry for a check written to petty cash for $100.00. Many times you will need cash around to use for various things. In order to keep close tabs on where the cash goes once it is put into the petty cash box, you need a Petty Cash Journal. You will find that it is much better to run all your income through your checking account. That way, every time you put money into your petty cash box, you will have a sure record of where the money came from. In other words, don't cash one of your mill checks and put it into the petty cash box. This will create many problems for your accountant.

Making Journal Entries. In Figure 20, you will notice that the petty cash journal pages are similar to the other journals in our system. You will also notice that our first entry was the $100.00 that we took from the checking account. Each time you put money into the petty cash fund, you will enter it in this manner. Expense entries are made in the following manner:

- The *date* is entered first.
- The *item purchased* or what the money was spent on is entered next.
- The *amount* paid out is entered.
- The balance after subtracting the expense is entered.

Each of the other entries is made in the same manner.

At month's end you will total column 1 and carry your balance forward to the next month (if there is money left in the box). Each month is done in the same manner. In this way you have a month-by-month summary of all of your petty cash expenditures.

TRIANGLE M LOGGING Co.

PETTY CASH JOURNAL For the year 19 80 MONTH OF OCTOBER

DATE		ITEM PURCHASED OR PAID OUT	1 AMOUNT	2 CASH INPUTS	3 BALANCE
10	22	FROM CHECKING ACCOUNT		100 00	100 00
10	23	PAPER TOWELS FOR OFFICE	1 18		98 82
10	24	PERMITS TO MOVE SHOVEL	6 00		92 82
10	27	SOFT DRINKS FOR OFFICE	7 50		85 32
10	29	LIGHT BULBS FOR SHOP	10 30		75 02
10	30	PERMITS TO MOVE TOWER	6 00		69 02
10	31	LUNCH FOR PRE-WORK CONF.	15 00		54 02
		OCT. TOTAL EXPENDITURES	45 98	BAL. FOR	54 02

Figure 20. Sample ledger sheet—petty cash journal

Usually, when a company uses a petty cash system, one person is authorized to approve such expenditures. Receipts should be kept and if you think it necessary, petty cash vouchers can be used (see Figure 21). One copy of the voucher should be given to the employee receiving payment and one copy placed in the petty cash box. Completing the form is self-explanatory.

SHOULD YOU USE PURCHASE ORDERS?

The use of purchase orders is somewhat controversial. Robert Townsend says in *Up the Organization,* ". . . don't underestimate the morale value of letting your people 'waste' some money. If you must, have a one-man 'buying department'. . . ."

Townsend is of course speaking of a purchasing department, not purchase orders. But what he says about "waste" can be true if you have no controls at all on

PETTY CASH VOUCHER

DATE 9/10/81 | $ 2.58

ITEM PURCHASED:

PAPER CUPS & MENDING TAPE (RECEIPTS ATTACHED)

CHARGE TO CODE: 539

SIGNED: Pauletta Carter

Figure 21. Petty cash voucher

PURCHASE ORDER

PLEASE PUT THIS NUMBER ON ALL INVOICES-PACKAGES, ETC.

NO. **1234**

TO: SOUTHERN PACIFIC AUTO SUPPLY

ADDRESS: 1342 SOUTH MAIN — DATE: 6/15/81

SHIP TO: TRIANGLE M LOGGING CO. — JOB NO. 12

	QUANTITY	PLEASE SUPPLY ITEMS LISTED:	UNIT PRICE	TOTAL PRICE
1	8	AC SPARK PLUGS # HB-28	1.10	8.80
2	5 FT.	BLACK HEATER HOSE	.50 FT.	2.50
3	38	1/4" X 2" BOLTS	.50	19.00
4	38	1/4" HEX NUTS	.20	7.60
5	38	1/4" LOCK WASHERS	.09	3.42
6				
7		TOTAL THIS ORDER		$41.32
8				
9				
0				

DATE REQUIRED: 6/15/81 — HOW TO SHIP: WILL PICK UP — PREPARED BY: LOGAN FRANK

TERMS: NET 30 — APPROVED BY: Raymond Bradbury

Figure 22. Simple purchase order

your people. If you don't use purchase orders, you really don't have any control over spending. So, drop the department concept, but not the purchase order.

A purchase order system does not have to be complicated. In Figure 22, you will see a simple purchase order form that can be filled out quite easily. The controlling device is the serialized number in the upper right-hand corner. If each member of your company has to have a purchase order number before he makes a purchase, he will think twice about making the purchase. If it is really important, he will call the office for a number and make his purchase; if it is not important, he will usually control his urge to buy and wait until the purchase is really necessary. In this way, you will not be buying wash mitts, auto polish, spark plugs and so forth, for every employee to take home.

PURCHASE ORDER CONTROL SHEET PAGE 2

P.O. NO.	WRITTEN TO	REQUEST OR JOB NO.	CHARGE CODE	DATE OF ORDER	AMOUNT OF ORDER	VENDOR INVOICE NO.	PMT. DATE
1230	OAKRIDGE LOGGING	12	505	5/1/81	$1500.00	2438	6/1/81
1231	CHAPMAN BROS.	OFFICE	539	5/2/81	$35.80	03578	6/2/81
1232	LARRY'S LUMBER	12	505	5/3/81	$243.00	7351	6/3/81
1233	HOWARD COOPER	12	509	5/5/81	$850.00	00531	6/5/81
1234	SO. PAC. AUTO SUPPLY	12	523	6/15/81	$41.32	0834	7/15/81

Figure 23. Purchase order control sheet

One day I asked one of my suppliers who could charge against my account. He told me that anyone walking in off the street could charge against my account if I didn't give his store a list of authorized people or use a purchase order system. We started using purchase orders the very next day.

In order to keep the purchase orders in the proper sequence, we used the simple purchase order control sheet shown in Figure 23. As the numbers are used, they are logged onto the sheet with the other pertinent information shown. These forms also help you track parts and other supply expenditures by individual job number.

ANALYSIS OF MONTHLY EXPENSES, ALL JOBS

If you wish to see a total month's evaluation, use the form shown in Figure 24. It is simple to compile. Just take the totals from your cash disbursements journal for

MONTHLY COST ANALYSIS FORM FOR THE MONTH OF OCTOBER, 19 80

EXPENSE SUMMARY BY CODE NUMBER		TOTALS
501	Timber Purchases ($ paid on volume)	
502	Logging Subcontractors	1000.00
503	Logging Equipment Rental	1407.60
504	Logging Equipment Maintenance & Repair	1287.20
505	Logging Supplies-Wire Rope & Rigging	4977.80
506	Logging Saw Maintenance, Repair & Purchase	252.50
507	Scaling Fees	54.00
508	Road Building, Fire Trails, & Clean-up	4800.00
509	Miscellaneous Logging Expense	1880.00
510	Equipment Relocation Costs	1200.00
511	Equipment Lease Costs	3500.00
512	Equipment Purchases	6400.00
513	Contract Cutters	16350.00
521	Trucking Sub-contractors (Gypo's)	17850.00
522	Trucking Equipment Rental	1898.00
523	Trucking Equipment Maintenance & Repairs	3585.00
524	Licenses, Permits, & PUC Plates	320.00
525	Fines & Citations	57.00
526	Miscellaneous Trucking Expenses	1015.00
527	Tire Expenses	1950.00
528	Petro-products, (gasoline, diesel, oil, etc.)	4623.20
529	Radio & Communications	385.80
536	Cleaning, Maintenance, & Sanitation Service	120.00
537	Utilities	192.50
538	Shop Supplies	1350.80
539	Office Supplies	285.80
540	Dues & Subscriptions (Trade journals)	48.00
541	Postage & Shipping Charges	90.00
542	Charitable Contributions	500.00
543	Training & Certification	425.00
544	Entertainment & Travel	1287.80
545	Advertising & Public Relations	342.40
	Page sub-totals	79135.40

Figure 24. Monthly cost analysis form—October, 1980

Page two. Monthly Cost Analysis

Code	Account	Amount
546	Miscellaneous Office Expenses	200.00
547	Legal & Professional Fees	2780.00
548	Payroll-Officers (Monthly)	2400.00
549	Payroll-Hourly	5067.80
550	Payroll Taxes (FICA, Federal, State, Etc.)	3250.00
551	Personal Property Taxes	3800.00
552	Real Property Taxes	1500.00
553	Income Taxes	300.00
554	Other Taxes	2950.00
555	Insurance (Worker's Comp., Medical, etc.)	2800.00
556	Interest Paid	2400.00
557	Reporting Penalties & Fines	32.60
558	Bank Charges	3.58
	Grand Total both Pages	106619.38

Total Income all Sources	110318.00
Total Expenditures all Codes	106619.38
Total Profit for period	3698.62
Total Loss for period	( )

FINAL ANALYSIS: ____________________

Report Prepared By: Ronald R. Macklin Date 11/3/80

Signature ____________________

Figure 24. *Continued*

each coded account and take the total income figures from your cash receipts journal. Fill in the blanks as shown in the figure and see what happened to your company this month.

SUMMARY

In this chapter we did not endeavor to teach you all about bookkeeping. We have only covered these concepts directly related to the Timber Harvesting System explained in Chapter 8. The instructions necessary for an entire set of books for your company are far beyond the scope of this small book.

Your accountant can help you develop all the other things you need for a full-blown bookkeeping system.

Remember: *accuracy through detail.*

REFERENCES

Davidson, Sidney, and Roman L. Weil. 1977. *Handbook of Modern Accounting.* 2d ed. New York: McGraw-Hill.

CHAPTER 8

Establishing Controls and a Cost Analysis System

Usually, the first question to come to mind when you think about getting control of your business is: "Where do I begin?" Once you find the answer to that question, you then want to know how to go about gathering and using information and data to help run your business in a successful manner.

If you are just starting out in the logging business (or any other business), you are in a fortunate position as far as record keeping is concerned. You can begin paying attention to details from the very start. You can begin at the beginning.

But what about the guy who has been in business for years? Where does he begin? Unfortunately, the fellow must begin in the middle of things and that is not the easiest place to start. For the man already in business, the best way to implement a new control system is to pick a date, preferably the first day of a new month, and start.

Begin by collecting everything you can get your hands on—every invoice, every receipt, every cancelled check, every time card and all the other data you can get from the previous month's activity. This is the information and data needed for you to function as a true business executive and manager who will plan, organize, motivate, control and coordinate.

Now we must begin to tell you how to use all this information, how to assemble it so that it makes sense and how it can tell things about your business you never dreamed existed. You should then be able to tell, all the time, whether your business is a success or a failure.

In order to be competitive, a logger must know exactly how much it costs to log. His prices should then reflect that cost and obviously a modest and fair profit (a word almost unknown to many loggers today).

As you must know by now, in today's world, breaking even is not good enough. Remember, *you are not in business for the experience.* You are in business to *make money!*

GATHERING AND USING INFORMATION

In setting up a cost control system you must have as much depth in your analysis as is possible. The depth of the analysis will depend on three factors:

- *Time.* The time necessary to do the analysis.
- *Tools.* The tools needed to accomplish the analysis.
- *Funds.* The money necessary to pay for the analysis.

In this book, we cannot give you time or money, but can provide the tools you need to put it all together. We can guarantee, right now, that you will be able to justify every hour and every dollar spent on this project. You will also be able to add this administrative cost to logging prices just as you do every other expense.

Even in logging, a manager is only as good as the information he has and the facts available to him for making management decisions. To make these decisions, you need *cold, hard, calculated* facts. This cost analysis program will help find the answers to such questions as:

- What is my overhead cost?
- What equipment should I buy?
- What equipment should I repair or rebuild?
- How many men do I need to do a particular job?
- What is my profit and loss picture?
- Is my current equipment doing the job?

You will also receive the answers to many other questions about your business.

We will approach the problem of establishing a cost control system in an analytical way: *one step at a time.* Remember, every journey begins with a single step, and you have already taken the first step when you started reading this book. This step-by-step analytical approach has worked for others in every aspect of setting up a cost control system.

THE ACTUAL PROGRAM

The program can go several ways. You can hire someone to come in and do it, you can use existing people or you could even do it yourself. Regardless of the method, you must spend the time and attention necessary to complete the project. All of this may prove to you that you have been working for several years just breaking even and not for the profit-making, growth-oriented business you might desire.

Assigning and Using Job Numbers

To begin the orderly gathering of information, the information must have a place to live while it is being used and then a place to remain as it becomes the historical background of your business. We begin the control system with the first in a series of forms designed to gather the information needed for the system. Form No. 1, Job Site Control Numbers, is the foundation of the entire control system.

Form No. 1 contains the following things:

1. *Job number.* The numbering sequence can begin anywhere. We have found that using the year and a sequential number works very well. So we start with 81-1, 81-2, 81-3 and so on.
2. *Job name.* This is the name used by the Forest Service or other agency for the sale or the geographic location name, i.e., Carpenter Bypass, Cougar Creek, and so forth.
3. *Mill.* Who are you doing the job for? Identify the mill with whom you have the contract.

Form No. 1. Job site control numbers

JOB SITE CONTROL NUMBERS

JOB NO.	JOB NAME	MILL	START	FINISH	BRAND	PRICE M	OTHER
81-1	LORANE CARPENTER BYPASS	BANGLE VENEER	2/28/81	5/13/81	C	85.00	
2	TAPER SALVAGE	AMAZON WOOD	3/24/81	7/14/81	A1	75.00	77433
3	TILLER FALLS	EVERGRN PLYWOOD	3/18/81	5/24/81	EP	60.00	
4	BILLS SALVAGE	EVERGRN PLYWOOD	3/3/81	4/13/81	EP3	75.00	
5	DOMINO	AMAZON WOOD	7/15/81	10/13/81	A4	85.00	77501
6	GATE CREEK	BANGLE VENEER	7/20/81	12/14/81	C	55.00	
7	WHIRL	AMAZON WOOD	6/20/81	12/5/81	A3	90.00	77527
8	SANDY CREEK	BANGLE VENEER	5/14/81	9/10/81	C2	65.00	
9	SLAP DOWN	AMAZON WOOD	6/12/81	10/10/81	A5	85.00	77602
10	CUPIT MARY	EVERGRN PLYWOOD	8/4/81	12/15/81	EP4	75.00	
(1)	(2)	(3)	(4)	(5)	(6)	(7)	(8)

4. *Date the job was started.*
5. *Date the job was finished.*
6. *Brand being used to identify the logs.*
7. *Price per thousand board feet being paid.*
8. *Any other information needed* to identify every aspect of the job. (Some mills use code numbers of their own to identify jobs, etc.)

This form becomes a permanent part of your record system.

Making up the Job Folder

After the number is assigned to a job, the next thing to do is to make up a job folder (Figure 25). This folder houses all the information collected on the job. This is where the information will live while it is being used, and the place where it will remain as history. The more data accumulated, the more accurate and complete the analysis will be.

As seen in Figure 25, the job folder displays the job number, name of the job and the brand. The start and finish dates should be placed on the outside of the folder as a reference. All the other system-related forms will be covered as we continue to put together a cost control system.

THE OTHER FORMS IN THE CONTROL SYSTEM

The forms in this system are designed around the basic elements of a timber harvesting system. This basic system in sequential order can be seen in Figure 26.

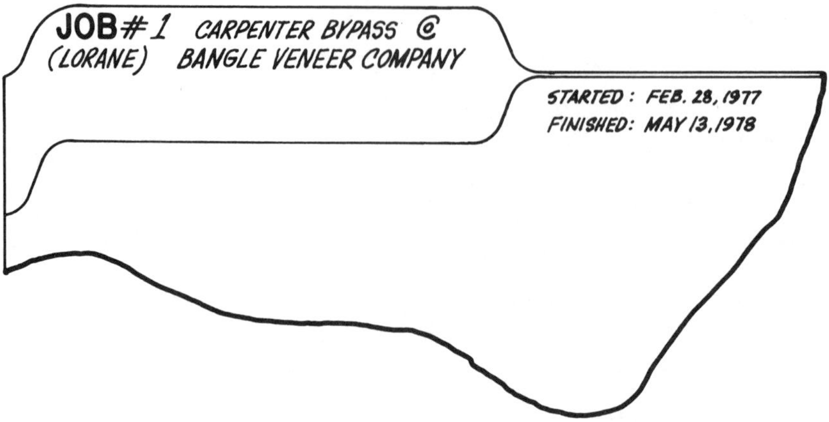

Figure 25. The job folder

Some may feel the forms in the system are overly complex. For some operations they may be, but remember that the system was designed for a company having no records at all. You may wish to eliminate some of the forms in the system. You may wish to add some of your own. Regardless, it is good to review the entire timber harvesting system to make sure you are gathering information concerning every step of the operation from start to finish. There are 19 elements shown in Figure 26. Each numbered element is identified as follows:

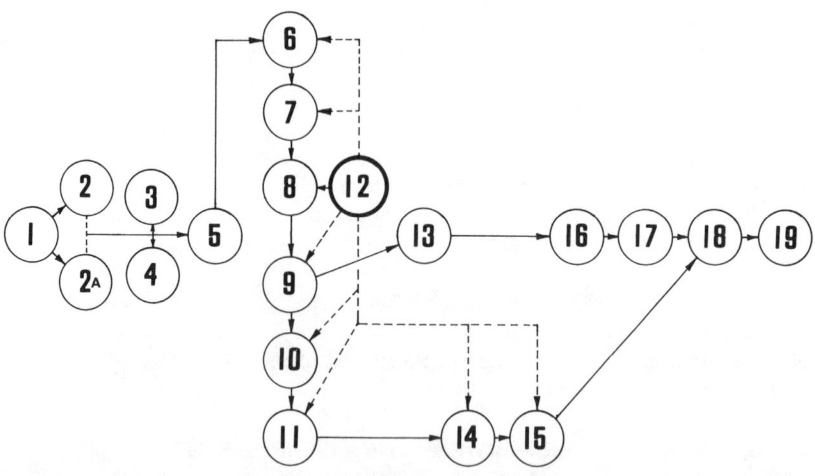

Figure 26. Timber harvesting flow diagram

1. Bidding the sale with the Forest Service, BLM, mill, etc.
2. Buying the sale, or . . .

2a. Negotiating the contract with the mill.

3. Financial planning and purchasing.
4. Job planning; personnel and equipment selection.
5. Equipment transport and placement.
6. Felling, measuring, bucking, limbing and perhaps topping.
7. Yarding and decking (would also include tree pulling).
8. Sorting, marking, branding and loading.
9. Hauling and scaling (or weighing).
10. Equipment maintenance, repair and related items.
11. Road construction.
12. Constant monitoring and financial analysis of personnel and equipment.
13. Unloading.
14. Cleanup and slash disposal (for which you are responsible).
15. Reforestation (for which you are responsible).
16. Final inspection by the contracting agency.
17. Equipment removal, transport and relocation (same as No. 5 above).
18. Close out job.
19. Final financial analysis.

All 19 elements combine to make up a total timber harvesting system.

WHAT WE NEED TO KNOW ABOUT EACH ELEMENT

Elements 1, 2, 2a, 3, 4, 12 and 19 are extremely important to the financial success of the overall operation. Element 7 would also accommodate the use of an elaborate suspension system, if one must be used. Perhaps at times you might have to add an element or take some away, depending on the complexity of the operation.

Each element mentioned is important to the overall success of your business. There are special things we want to know about each of the elements. Let's review them and cover the forms designed for each element.

Elements 1 through 4: Management Decision-Making Functions

During the course of the book, we will suggest ways to accumulate data and information to help with these four elements. You should never be afraid to lose a bid if you know the price is too low to produce a profitable or breakeven operation.

Element 5: Transport and Placement of Equipment

We must know all costs for permits, fees, transportation and personnel associated with each equipment move. Form No. 2, the Job Relocation and Equipment Move form, will provide this input.

Many pieces of equipment can be transported with a lowboy. Some are self-propelled, and some must be subcontracted due to excessive size, weight or the fact

Form No. 2. Job relocation and equipment move

JOB RELOCATION AND EQUIPMENT MOVE (1) Date 4/18/81 Job No. 12 (2)

Moved From (3) QUARTZ CREEK

Moved To (4) COUGAR CREEK

Distance in Miles 40 (5) Weather Conditions CLOUDY & WET (6)

Permits, Fees, Tags, Associated Costs STATE & COUNTY $6.00 (7)

ITEMS MOVED OR INVOLVED IN MOVE (number of each) Hrs. 8 (8)

(9) Tractors [ ] Skidders [/] Tower [/] Grader [ ] Firetruck [ ] Pickup [/]

Crew Bus [ ] Misc. Truck [ ] Misc. Equip. [X] Saws [2]

Identification Numbers 4-57-233 & 7-59-234 (10)

PICKUP # 44

(11) Rigging Number of People Involved: Supervisory [/] Hourly [2] (12)

Remarks: SKIDDER MOVE SUB-CONTRACTED INVOICE # 2436

(13) SKIDDER ON LOWBOY, TOWER MOVED WITH #7 KENWORTH

that you must move two pieces of equipment at the same time and only have one lowboy available.

Form No. 2 is completed in the following manner:

1. Date the equipment is to be moved.
2. Job number assigned on the master log sheet (Form No. 1).
3. Location of the equipment at the time of the move.
4. Location to which the equipment is to be moved.
5. Total distance of the move.
6. Weather conditions prevailing at the time of the move, i.e., clear, rainy, snowing, and so forth.
7. Permits needed and the cost of the permits. Fees charged by subcontractors.
8. Hours involved in moving the equipment.
9. Place an X in the appropriate box indicating which equipment is being moved.
10. Equipment identification numbers, control numbers or theft and vandalism number.
11. Rigging being moved, if any.
12. Number of people involved in the move, supervisory and hourly.
13. Any pertinent remarks that would be helpful in describing the relocation/move or that would further clarify the costs. (Attach invoices, if possible.)

Remember to file the completed form along with all related information in the project folder made up to store the job information.

Element 6: Felling, Measuring, Bucking, Limbing and Topping

We will want to know all costs associated with these activities. If your own employees take care of these operations, it is a simple matter to check the weekly time cards and time analysis sheets to determine the hours expended. The volume is determined as the timber is transported to the mill. As you know, each load has a scale ticket and the ticket summary sheets provided by the scaling bureau will give you the exact volume coming out of each sale.

If you are using subcontractors, getting the information may be a little more difficult. If you are paying the contractor by the thousand, it will be easier to get the information. If you are paying by the man-day, you will have to insist that the subcontractor provide detailed information concerning his activity on each job.

As yet we have developed no specific forms for this element. If you require specific data that cannot be provided by other forms in the system, you can develop those forms for your own operation.

Element 7: Yarding and Decking

Once again, we need to know all the costs associated with these activities. We want to know the number of turns per shift and the number of logs per turn. We will also want to know how much the equipment is costing us to operate and the number of operator hours involved.

To record this information accurately with the minimum amount of errors, we have developed two forms to be used in conjunction with the weekly time cards.

The first of these is Form No. 3, the Daily Yarding Report.

This form is used by all skidder, tower or slackline operators on a daily basis. The form is completed in the following way:

1. Job number assigned from the master log sheet.
2. Date the work is accomplished.
3. Full name of the operator.
4. Job location or map location coordinates.
5. The number of logs for each turn. (The form accommodates 60 turns.)
6. Total number of turns each day.
7. Total number of logs for the day.
8. Yarding method used—i.e., skidder, high-lead tower, slackline, and so forth.

Notice that Form No. 3 provides the total productivity for each side, each day. Coupled with Form No. 4, the Equipment Use form, and Form No. 5, the Weekly Time Card, you can evaluate not only equipment productivity but the productivity of your personnel at the same time.

The Equipment Use form (Form No. 4) is used by all equipment operators (with the exception of log truck drivers and pickup and crummy drivers, who have other control forms). Form No. 4 contains the following information:

Form No. 3. Daily yarding report

DAILY YARDING REPORT JOB NO. 12 (1)

(2) DATE 5/12/81 OPERATOR LARRY LAMKIN (3) JOB SITE COUGAR CREEK (4)

(5)

TURN	LOGS	TURN	LOGS	TURN	LOGS	TURN	LOGS	TURN	LOGS	TURN	LOGS	
1	3	11	3	21	4	31	3	41	3	51		
2	4	12	3	22	3	32	3	42	4	52		
3	3	13	3	23	2	33	3	43	3	53		
4	3	14	4	24	2	34	4	44	3	54		
5	3	15	2	25	3	35	3	45	3	55		
6	3	16	3	26	3	36	3	46	2	56		
7	4	17	3	27	3	37	3	47	3	57		
8	2	18	3	28	3	38	3	48		58		
9	3	19	3	29	3	39	3	49		59		
10	3	20	3	30	3	40	2	50		60		

(6) TOTAL TURNS: 47 | TOTAL LOGS: 141 (7) | YARDING METHOD: HIGH LEAD TOWER (8)

1. Job number as assigned on the master log sheet.
2. Odometer reading. If the equipment is self-propelled, as in the case of the FMC skidder, log the speedometer or wheel odometer reading at the beginning and ending of each day or week. (In the sample form, the odometer is broken, which indicates a need for maintenance or repair.)
3. Operator's name.
4. Job location or map designation coordinates.
5. Date the week began.
6. Description of equipment or type of equipment.
7. Number of actual hours equipment was in operation. (Remember, this will not always match the employee's time card for the same period.)
8. Number of gallons consumed (with the weekly total) if gasoline powered.
9. Number of gallons consumed (with the weekly total) if diesel powered.
10. Number of quarts or gallons of oil consumed (with the weekly total).
11. Pounds of lubricant (or number of tubes) used (and the weekly total).
12. Number of hours the machine was not productive during the working day due to equipment malfunction. (This accounts for the difference between the equipment utilization time and the worker's actual time on his weekly time card.)
13. Note problems needing attention of maintenance personnel.
14. Suggestions or things to do to solve the problems encountered.

The Weekly Time Card (Form No. 5) is the record of all time spent each day by every employee. This form is used in conjunction with several other forms in the

Form No. 4. Equipment use (weekly)

EQUIPMENT USE (WEEKLY)

(1) JOB NO. 12

(2) MILEAGE BEGIN BROKEN END

(3) OPERATOR NAME JEFFREY DOE DATE 5/21/81 (5)

(4) JOB LOCATION COUGAR CREEK EQUIP. TYPE FMC SKIDDER (6)

	DAY OF THE WEEK	MON	TUES	WED	THUR	FRI	SAT	SUN	TOTAL FOR WEEK
(7)	OPERATING HOURS	8	7½	8½	8	6½	1½		40.0 HRS
(8)	GAS IN GALLONS	/	/	/	/	/	/	/	0
(9)	DIESEL IN GALLONS	48		39		49			136 GALS.
(10)	QUARTS OF OIL	4	4			4			12 QTS.
(11)	LUBE			3					3 LBS
(12)	DOWN TIME		½			1½			2.0 HRS

(13) MAINT. OR REPAIRS MINOR ADJUSTMENT TO BRAKE SYSTEM

REMARKS SUGGEST DETAILED CK OF BRAKE SYSTEM TO ASSURE

(14) SATISFACTORY OPERATION.

(FIX CABLE OR WHEEL ODOMETER) JDD

system. Each will be tied in at the appropriate time in our coverage of the total system.

Every crew (or every employee not on a crew) is required to keep a daily record of the total hours worked each day. Even monthly people should be required to do

Form No. 5. Weekly time card

WEEKLY TIME CARD

Job No. 12 (1)

Truck No. (3) Job Location COUGAR CREEK (2)

EMPLOYEE'S NAME (4) NO. (5) WEEK ENDING 5/10 (6) 19 81 (7)

CREW NAMES	KIND OF WORK DONE	S	M	T	W	T	F	S	HRS.	RATE	STATUS CHG.
LAMPKIN	YARDER OPR. (9)		8	8½	9	8	9	4	46½		
DOE (8)	FMC OPR.		8	7½	8½	8	6½	1½	40		(13)
SMITH	CHOKER SETTER	(10)	8	8½	9	8	9	4	46½		
JONES	CHOKER SETTER		8	8½	9	8	9	4	46½	(12)	
MARTINEZ	CHASER		8	8½	9	8	9	4	46½		
SOUTHWARD	LOADER OPR.		8	8	8	8	9	4	45		
MACKLIN	HOOK TENDER		8	9	9	8	9½	4	47½		
									(11)		
	Total Regular Time	(14)	56	55½	56	56	54½		278		
	Total Overtime		(15)	3	5½		6½	25½	40½		

Form No. 6. Job site equipment utilization summary

JOB SITE EQUIPMENT UTILIZATION SUMMARY

(1) JOB NO. *12* LOCATION *Cougar Creek* (2) BRAND *A4* (3) FOR *Amazon Wood Products* (4)

19 *81* (5) MONTH OF *May* (6)

EQUIP.	EQUIPMENT ITEM	M	T	W	T	1 F		4 M	5 T	6 W	7 T	8 F		11 M	12 T	13 W	14 T	15 F		18 M	19 T	20 W	21 T	22 F		25 M	26 T	27 W	28 T	29 F		
(7) *FMC*	DIESEL				(8)		(9)		45			40	85		46			44	90			45			45			43			43	263
	OIL									5			5			20			20				3		3		2				2	30
	LUBE					2	2			2			2			2			2		2				2			2			2	10
	DOWN TIME													½	½	1	½	½	3													3
	RELOCATION																															(10)
	OPERATOR HRS.				8	8	16	8	8½	9	8	12	45½	7½	8	8	7½	12½	43½	8	8	8	8	8	40	9	8½	9	9½	10	46	191
Skagit BU80	DIESEL								460				460				320		320			240			240			400			400	1420
	OIL								6		6		12	6		6			12		6				6		6			6	12	42
	LUBE								1		4		5			14			14		4			4	8			4		4	8	35
	DOWN TIME					½	½								½		½		1			½			½				1½		1½	3
	RELOCATION																															
	OPERATOR HRS.				8	8	16	8	8	8	8	8	40	8	7½	8	7½	12	43	8	8	8	8	8	40	9	8	8½	9	9	43½	182½
	DIESEL																															
	OIL																															
	LUBE																															
	DOWN TIME																															
	RELOCATION																															
	OPERATOR HRS.																															
	DIESEL																															
	OIL																															
	LUBE																															
	DOWN TIME																															
	RELOCATION																															
	OPERATOR HRS.																															

this. They would simply report 40 hours each week and would sign the time card to that effect.

The crew chief, or hook-tender, should be responsible for each crew's time card. Individuals working alone will fill out their own cards. All work time, both regular and overtime, is accounted for.

The time card (Form No. 5) contains the following information:

1. Job number assigned from the master log sheet.
2. Location of the job, by name or by designation.
3. If the time card is used by a driver, his truck number would appear here.
4. Employee's full name.
5. Employee's number (if you have assigned one).
6. The day ending each work week. This can be Saturday or Sunday, as you choose. The actual calendar date is entered here.
7. Current year, fiscal or calendar.
8. Names of individual crew members as entered by the crew-chief or hook-tender.
9. Kind of work being done by each member of the crew or by each employee.
10. Hours worked each day by each member of the crew or by each individual.
11. Total number of hours worked that week by each employee.
12. Rate of pay for each person on the card. (This is added after the card is turned into the office.)
13. Any change in status for an employee—i.e., a raise, fired, laid off, and so forth.
14. Total number of hours for each day, all employees.
15. Total number of hours overtime each day, all employees.

Time cards should be turned in on a weekly basis and should be retained in your payroll records file. Job site analysis sheets are kept in each job folder for each month's activity. This leads us to two other important forms that should be mentioned at this time. These two forms work in harmony with the Weekly Time Card and the Equipment Use form. These forms are used simply to give you the month's activity at a glance for equipment use and employee time.

Form No. 6, the Job Site Equipment Utilization Summary, is used in conjunction with Form No. 4. As mentioned, it simply provides you with the month's activity at a glance. You will transfer the data in the following way:

1. Job number as assigned from the master log sheet.
2. Location of the job.
3. Brand being used to identify the logs.
4. Mill or agency you are logging for.
5. Year of the activity.
6. Month of the activity.
7. Name of the equipment being used.
8. Daily totals for each item.
9. Weekly totals for each item.
10. Monthly totals for each item.

Form No. 7. Job site manpower utilization summary

JOB SITE MANPOWER UTILIZATION SUMMARY

JOB NO. 12 (1) LOCATION COUGAR CREEK (2) BRAND A4 (3) FOR AMAZON WOOD PRODUCTS (4)

19 81 MONTH OF MAY

EMPLOYEE NAME	M	T	W	T	1 F		4 M	5 T	6 W	7 T	8 F		11 M	12 T	13 W	14 T	15 F		18 M	19 T	20 W	21 T	22 F		25 M	26 T	27 W	28 T	29 F		
(5) LARRY LAMPKIN				(6)	8	8	8	8½	9	8	13	46½	8	8½	9	8	13	46½	10	9	8	9	9	45	8	8	8	8	8	40	(9)
JEFF DOE					8	8	8	7½	8½	8	8	40	8	7½	8½	8	8	40	10	9	8	9	9	45	8	8	8	8	8	40	
TERRY SMITH					8	8	8	8½	9	8	13	46½	8	8½	9	8	13	46½	10	9	8	9	9	45	8	8	8	8	8	40	
DALE JONES					8	8	8	8½	9	8	13	46½	8	8½	9	8	13	46½	10	9	8	9	9	45	8	8	8	8	8	40	
PIO MARTINEZ					8	8	8	8½	9	8	13	46½	8	8½	9	8	13	46½	10	9	8	9	9	45	8	8	8	8	8	40	
SID SOUTHWARD					8	8	8	8	8	8	13	45	8	8	8	8	13	45	10	9	8	9	9	45	8	8	8	8	8	40	
ARN MACKLIN					8	8	8	9	9	8	13½	47½	8	9	9	8	13½	47½	10	9	8½	9½	9½	46½	8½	8½	8½	8½	8½	42½	
						(7)																									
DAILY TOTALS				(8)	56		56	58½	61½	56	86½		56	58½	61½	56	86½		70	63	56½	63½	63½		56½	56½	56½	56½	56½		
WEEKLY TOTALS						56						318½						318½						316½						282½	
						(10)																									
MONTHLY TOTAL																															1292
																															(11)

After completing this form, the daily records can be filed away in the appropriate job folder as historical data.

Form No. 7, the Job Site Manpower Utilization Summary, is used in conjunction with Form No. 5. As mentioned, it simply provides the month's activity at a glance. You will transfer the data in the following way:

1. Job number as assigned from the master log sheet.
2. Location of the job.
3. Brand being used to identify the logs.
4. Mill or agency you are logging for.
5. Employee's name.
6. Employee's hours each day.
7. Total hours for each week (individually).
8. Total hours for each day (collectively).
9. Total hours for the month (individually).
10. Total hours for each week (collectively).
11. Total hours for each month (collectively).

After completing this form the weekly time cards can be filed away in the company payroll records section.

Since rigging is a major expense associated with Element 7, you will want to keep track of all related costs. Not only will you want to keep track of where the rigging is located, but you will also want to charge the cost to each specific job.

To help accomplish this the Rigging Control Sign-out Sheet, Form No. 8, was developed. For this particular form to work, the rigging must be kept in a controlled location.

Form No. 8. Rigging control sign-out sheet

RIGGING CONTROL SIGN-OUT SHEET

FOR MONTH OF MAY 1981 (1)

QTY. NO.	ITEMS REMOVED	TAKEN BY (NAME)	JOB LOCATION	DATE REMOVED
2	7/8"x30' HILEAD CHOKERS	LARRY LAMPKIN	COUGAR CREEK	5/8/81
(2) 1	TAG LINE (3)	GREG BROWN (4)	SWITCH BACK (5)	5/9/81 (6)
2	5/8"x CAT CHOKERS	JOHN BAXTER	MINI BEAR	5/12/81
2	1"x30' CHOKERS	DAVE HARRIS	ELK MTN	5/13/81
1	LOADING LINE	DAN ROBERTS	WEST ANDY	5/15/81
3	7/8"x30' HILEAD CHOKERS	ROGER DAVIS	MT. JUNE	5/15/81
23	RAILROAD SPIKES	CLIFF MILLER	COUGAR CREEK	5/18/81

As the rigging is removed and taken to the various job sites, entries are made on the control sheet. A visual check is made each week to determine which items must be replaced in the centrally controlled area.

As each log sheet is filled, it is picked up and taken to the office, where a check is made against the actual invoices from the suppliers. Items do not vanish with this procedure and very seldom are we charged for something that we did not actually receive.

If rigging is destroyed, it is listed on Form No. 4 in the remarks column. It is then written off, being charged to the specific job that destroyed its usefulness.

Form No. 8 is completed in the following way:

1. Calendar month and year.
2. Number of items taken or returned.
3. Specific name and size of item.
4. Person physically removing or returning the item.
5. Job location where rigging is to be used or where it is returning from.
6. Specific date item was removed or returned.

At the same time an item is charged to a specific job number (by matching location with number), items returned can be credited back as they are returned.

Some items can be used continuously until such time as they are finally rendered useless or lost. The job destroying the item will be charged for the item's full replacement value.

When completed, the forms are filed in their own historical folder.

Element 8: Sorting, Marking, Branding and Loading

Throughout the system we are always looking for the total costs associated with each element of the timber harvesting system. You will basically use the same forms as for Element 7. Form Nos. 3, 4, 5, 6, 7 and 8 provide most of the information needed to evaluate this element.

In addition to the aforementioned forms, you will need to have Form No. 9, the Daily Load Report form. This form is used by all loader operators on a daily basis.

Form No. 9 is completed in the following way:

1. Job number assigned from the master log sheet.
2. Date that work is accomplished.
3. Full name of the operator.
4. Job location or map designation.
5. Destination of logs, or name of the mill or other facility where they are to be delivered.
6. Log truck used to haul logs, whether yours or that of a subcontractor.
7. Number of loads hauled by each truck. (As you see from the sample form, all trucks on this side are making 3 loads per day.)
8. The brand being used to identify the logs.
9. The total number of loads taken from the side on that day.

Form No. 9. Daily load report

DAILY LOAD REPORT

JOB NO. 12 (1)

(2) DATE 5/12/81 OPERATOR SID SOUTHWARD (3) JOB SITE COUGAR CRK (4)

DESTINATION	TRUCK HAULING	NUMBER OF LOADS	BRAND
(5) AMAZON W.P.	(6) TM #3	(7) 3	A4 (8)
	TM #4	3	A4
	TM #1	3	A4
BANGLE VENEER	TM #6	3	A4
	TM #2	3	A4
	O.T. PAINE	3	A4
EVERGRN PLY.	R. MONTGOMERY	3	A4
	TOTAL LOADS	(9) 21	

Remember, the number of logs and the volume of each load are contained on the scaling bureau summary sheets.

Form No. 10, the Monthly Load Summary Sheet, is used in conjunction with Form No. 9. This form is a monthly compilation of the information supplied on Form No. 9.

The data should be transferred in the following manner:

1. Job number assigned from the master log sheet.
2. Location of the job.
3. Brand being used to identify the logs.
4. Mill or agency the job is being done for.
5. Year of the activity.
6. Month of the activity.
7. Truck number or name of the subcontractor being used.
8. Number of loads hauled each day (individually).
9. Number of loads hauled each week (individually).
10. Number of loads hauled each month (individually).
11. Number of loads hauled each day (collectively).
12. Number of loads hauled each week (collectively).
13. Number of loads hauled each month (collectively).

With this information and other information from the system, you can determine the cost of this entire element. We will cover the correct procedure to determine total costs as the total system is developed.

Form No. 10. Load summary sheet (monthly)

LOAD SUMMARY SHEET (MONTHLY)

JOB NO. 12 (1) LOCATION COUGAR CREEK (2) BRAND A4 (3) FOR AMAZON WOOD PRODUCTS (4)

19 81 (5) MONTH OF MAY (6)

TRUCK	M	T	W	T	1 F	(9)	4 M	5 T	6 W	7 T	8 F		11 M	12 T	13 W	14 T	15 F		18 M	19 T	20 W	21 T	22 F		25 M	26 T	27 W	28 T	29 F		
TM #1 (7)					3	3	3	3	3	3	3	15	3	3	3	3	3	15	3	3	3	3	3	15	3	3	3	3	3	15	(10)
TM #2		(8)			3	3	3	3	3	3	3	15	3	3	3	3	3	15	3	3	3	3	3	15	3	3	3	3	3	15	
TM #3					3	3	3	3	3	3	3	15	3	3	3	3	3	15	3	3	3	3	3	15	3	3	3	3	3	15	
TM #4					3	3	3	3	3	2	2	13	3	3	3	3	3	15	3	3	3	3	3	15	3	3	3	3	3	15	
TM #6					3	3	3	3	3	2	2	13	3	3	3	3	3	15	3	3	3	3	3	15	2	3	3	3	3	14	
O.T. PAINE					3	3	3	3	3	2	2	13	2	3	3	3	3	14	2	3	3	3	3	14	2	3	3	3	3	14	
R. MONTGOMERY					3	3	2	3	3	2	2	12	3	3	3	3	3	15	2	3	3	3	3	14	2	3	3	3	3	14	
DAILY TOTALS			(11)		21		20	21	21	17	17		20	21	21	21	21		19	21	21	21	21		18	21	21	21	21		
WEEKLY TOTALS						21					(12)	96						104						103						102	
MONTHLY TOTAL																														(13)	426

Form No. 11. Log trucker's trip log

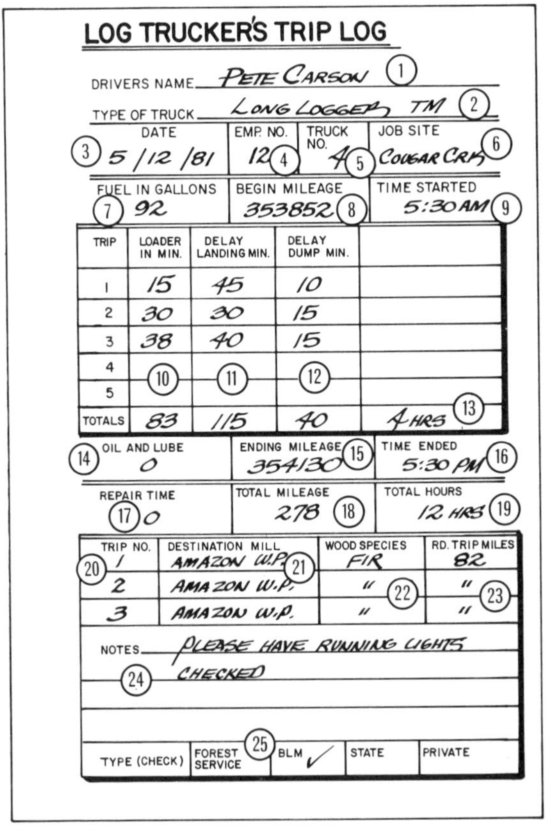

LOG TRUCKER'S TRIP LOG

DRIVERS NAME Pete Carson (1)

TYPE OF TRUCK Long Logger TM (2)

DATE	EMP. NO.	TRUCK NO.	JOB SITE
(3) 5/12/81	12 (4)	4 (5)	Cougar Crk (6)

FUEL IN GALLONS	BEGIN MILEAGE	TIME STARTED
(7) 92	353852 (8)	5:30 AM (9)

TRIP	LOADER IN MIN.	DELAY LANDING MIN.	DELAY DUMP MIN.	
1	15	45	10	
2	30	30	15	
3	38	40	15	
4	(10)	(11)	(12)	
5				
TOTALS	83	115	40	4 HRS (13)

OIL AND LUBE	ENDING MILEAGE	TIME ENDED
(14) 0	354130 (15)	5:30 PM (16)

REPAIR TIME	TOTAL MILEAGE	TOTAL HOURS
(17) 0	278 (18)	12 HRS (19)

TRIP NO.	DESTINATION MILL	WOOD SPECIES	RD. TRIP MILES
(20) 1	AMAZON W.P. (21)	FIR	82
2	AMAZON W.P.	" (22)	" (23)
3	AMAZON W.P.	"	"

NOTES Please have running lights checked (24)

TYPE (CHECK)	FOREST SERVICE (25)	BLM ✓	STATE	PRIVATE

Element 9: Hauling and Scaling (or Weighing)

In determining all the costs associated with Element 9, you will need to use Form Nos. 5, 7, 8 and 9, along with Form No. 11, the Log Trucker's Trip Log form.

This form should be used by all company drivers. The form provides all details necessary to evaluate driver and equipment performance. You will notice that Form No. 11 contains much detail and is filled out in the following way:

1. Driver's name.
2. Type of truck.
3. Date of the haul. (One form is used each day.)
4. Employee number.
5. Truck number.
6. Job site (location name).
7. Fuel used during the day in gallons.

8. Mileage at beginning of the day.
9. Time started in the morning.
10. Time spent in getting logs loaded.
11. Delay at landing before loading.
12. Delay at dump site while waiting to be unloaded.
13. Total time spent in delays.
14. Oil and lube used.
15. Mileage at end of the day.
16. Time ended in the evening.
17. Repair time.
18. Total mileage for the day.
19. Total hours for the day.
20. Trip number.
21. Destination of load.
22. Wood species hauled.
23. Round-trip mileage for each trip.
24. Any notes pertaining to day's activity.
25. Area from which logs were taken.

This record is very useful and in many locations required by state law. Combined with the other forms mentioned, it provides some revealing details for the cost analysis system.

Element 10: Equipment Maintenance, Repair and Related Items

Maintenance and repair is an important and expensive element, with many facets to be considered. The fine details and recommended procedures will be covered in Chapter 9.

There are, however, some related items that should be considered at this time.

In the course of gathering information, one needs data on the company-provided pickups and crummies, as well as fire trucks, fuel trucks, shop trucks and so on.

Pickups and Crummies. Regarding the performance of pickups and crummies, Form Nos. 12, 13 and 14 can prove useful:

The purpose of Form No. 12, Monthly Speedometer Readings form, is simply to determine the speedometer reading of each vehicle at the end of each month and who is using each of the trucks.

On the first working day of each month, this form is put on or near the gasoline pump where the trucks fill up each day.

Form No. 12 provides the following information:

1. Name of the driver.
2. Vehicle make and year.
3. Speedometer reading.
4. Company vehicle number.
5. Month ending.
6. Year.

Form No. 12. Monthly speedometer readings

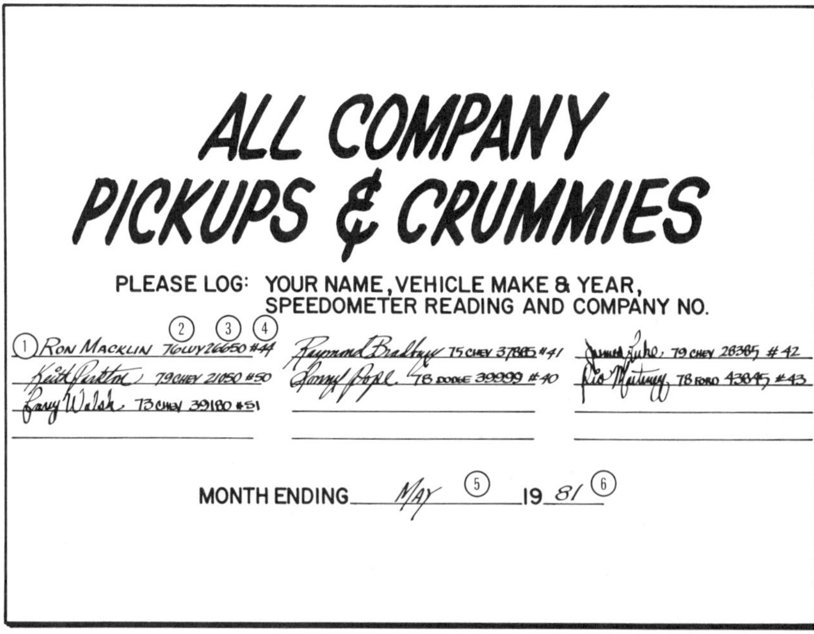
ALL COMPANY PICKUPS & CRUMMIES

PLEASE LOG: YOUR NAME, VEHICLE MAKE & YEAR, SPEEDOMETER READING AND COMPANY NO.

(1) Ron Macklin (2) 76 Chev (3) 26650 (4) #44

75 Chev 37885 #41

79 Chev 28385 #42

79 Chev 21050 #50

78 Dodge 39999 #40

78 Ford 43845 #43

73 Chev 39180 #51

MONTH ENDING May (5) 19 81 (6)

The form can be designed to accommodate as many vehicles as you require. By using this type of form, the drivers are required to make an entry only once each month rather than keep a detailed log like the log truck drivers.

After everyone has had an opportunity to fill in the information (usually two or three days), one of the office staff picks up the completed form. With the entries from the previous month's report, you can determine the total miles driven by each vehicle, each month.

Now that we have the mileage, we need to know how much fuel was consumed during the month by each of the vehicles on the list. This brings us to Form No. 13, Fuel Consumption (gasoline or diesel).

These forms can be put on a clipboard and posted somewhere near the fuel pumps. They remain there during the entire month. As each driver fills his tank or tanks during the month, he will make the appropriate entry. Since the date of the fill-up is not important, it need not be shown.

Each person whose name or vehicle number appears at the top of each column makes his own entries on the form.

Blanks 1 through 3 and item 6 are completed by the office. Item 5 is logged each time the vehicle is filled at the pump. The same procedure is used for either form, gas or diesel.

At month's end, the forms are picked up by the office staff. At this time, each column is totaled to determine the total gallons used by each driver.

Form No. 13. Fuel consumption form

FUEL CONSUMPTION FORM

(1) GASOLINE CONSUMPTION MONTH OF (2) MAY 19 81 (3)

TANKER	# 40	# 41	# 42	# 43	# 44	# 50	# 51	# (4)	OTHER
(5)	24.9	18.5	18.4	19.5	4.2	27.8	26.3		
	22.8	20.1	19.0	13.3	7.1	19.8	28.7		
	27.5	35.8	20.0	14.7	5.2	28.3	35.4		
	19.8	39.2	14.7	17.4	4.5	35.8	29.1		
	32.5	19.5	17.6	15.4	5.1	25.7	30.1		
	28.5	19.8	16.4	17.8	7.2	37.5	27.4		
	35.4	20.0	15.3	18.7		28.9	24.5		
	34.8	38.4	13.5	17.3	33.3	34.2	25.3		
	38.7	40.0	14.7			40.0	17.2		
	27.1	37.5		134.1	(6)				
		39.8	149.6			278	244		
	292	22.4							
		351							

FUEL CONSUMPTION FORM

(1) DIESEL CONSUMPTION MONTH OF (2) MAY 19 81 (3)

TANKER	# 40	# 41	# 42	# 43	# 45	# 48	# 50	# 5(4)	OTHER
100 (5)	85	100	73	58	50.0	50.0	87.5	67.3	25
850	70.2	88.1	92.7	95.8	50.0	47.3	63.2	68.9	12
785	83.5	65.8	20.0	20.1	70.0	75.0	25.0	57.9	14.7
100	97.3	86.5	99.8	30.2	40.7	74.0	75.8	90.1	50.0
950	50.2	87.4	87.9	72.1	39.8	84.4	24.1	92.4	25.7
875	48.5	83.2	35.5	87.7	38.9	75.5	65.2	98.7	26.8
975	68.8	99.2	88.3	97.8	67.5	89.8	95.8	63.4	88.7
300	88.7	98.7	74.1	75.3	98.7	87.5	23.5	99.2	25.0
500	47.3	75.3	83.2	84.1	99.8	20.0	24.1	40.0	30.0
873	73.5	57.5	97.8	59.5	40.3	21.0	97.8	65.4	47.8
685	50.3	59.3	38.9	55.6	75.6	97.2	98.9	74.8	78.3
783	50.4		39.8		57.8		97.5		12.5
500		901		736.2		721.7		818.1	6.8
485	813.7	(6)	831		729.1		778.4		27.8
790									
873									

Form No. 14. Mileage performance

MILEAGE PERFORMANCE

DATE (1) MAY 1981

VEH. NO.	DRIVERS NAME	BEGIN	END	MILES DRIVEN	GALLONS USED	MILES PER GALLON	VEHICLE
40	SONNY POPE	36781	39999	3218	292.0	11.0	78 DODGE
41	RAYMOND BRADBURY	35645	37885	2240	351.0	6.4	CHEV 75
42	JAMES LUKE	27185	28385	1200	149.6	8.0	CHEV 79
43	PIO MARTINEZ	42860	43845	985	134.1	7.3	FORD 78
44	RON MACKLIN	25809	26650	841	33.3	25.2	LUV 76
50	KEITH PARKTON	19250	21050	1800	278.0	6.5	CHEV 79
51	LARRY WALSH	37778	39180	1402	244.0	5.7	CHEV 73
(2)	(3)	(4)	(5)	(6)	(7)	(8)	(9)

Now that we have both the miles driven and the gallons used, we need to put all that information down on Form No. 14, the Mileage Performance form.

Make the following entries on the form:

1. Current month ending and year.
2. Vehicle number (assigned to each employee).
3. Driver's name.
4. Speedometer reading at beginning of the month (taken from Form No. 13 for previous month).
5. Speedometer reading at month's end (taken from Form No. 13 for current month).
6. Miles driven. (Subtract item 4 from item 5.)
7. Gallons consumed during month (taken from Form No. 14).
8. Miles per gallon. (Divide total miles driven by the total number of gallons consumed.)
9. Vehicle year and make.

If you wish to make a comparison of vehicle types and engine sizes to determine which vehicles are most economical, make a graph like the one in Figure 27.

Firetrucks, Fuel Trucks, Shop Trucks and Other Trucks. The information concerning other vehicles can be taken from Form No. 4, the Equipment Use form, items 2 and 8 or 9. They can be analyzed in the same way the pickups and crummies were analyzed and graphs can also be made showing related pieces of equipment for comparison purposes.

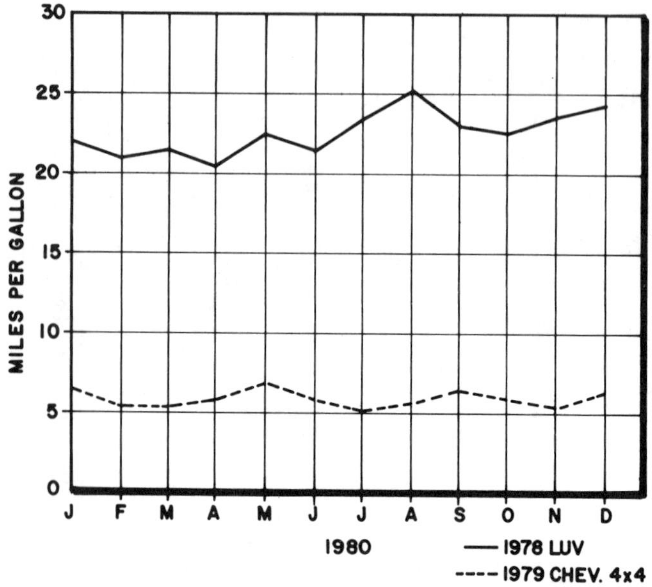

Figure 27. Comparison of vehicle economy

Element 11: Road Construction

Element 11 of the timber harvesting system can be analyzed by using information from forms already covered in other elements.

You should consult the following forms to gather your road-building costs and performance data: Form Nos. 4, 6 and 7.

If the road-building job is separate from a logging job, you would assign a job number to it just as you would a logging job.

You should keep the forms developed for each element of the road job in a road-building job folder. Your road job would be analyzed by the station or by the mile, rather than by board feet as in logging.

Element 12: Constant Monitoring and Financial Analysis

Element 12 involves the total system described to this point. By adding some additional input, we will then be ready to complete a Project Cost Analysis Report.

To make our input complete, we need two more things. The first is Form No. 15, a Cost of Operating Analysis form. One of these forms should be completed for each piece of equipment owned and operated by your company.

Form No. 15 contains the following information:

1. Date report is prepared.
2. Name of equipment being analyzed.
3. Monthly loan payment (if financed).

Form No. 15. Cost of operating analysis

COST OF OPERATING ANALYSIS

DATE 12/30/80 (1)

EQUIPMENT NAME BANTAM 266L LANCO CARRIER (2)

COST OF OWNERSHIP

	MONTHLY
LOAN PAYMENT	(3)
INTEREST ON LOAN	(4)
(5) DEPRECIATION	413.19
PROPERTY TAX	(6) 16.20
(7) LICENSE PLATES	25.00
INSURANCE	(8) 191.25
(9) RADIO	100.00
SUB-TOTAL	(10) $745.64

PRESENT VALUE (A)	$210,000.00
15 % FOR SALVAGE	31500.00 (B)
DEPRECIABLE VALUE (C)	178500.00
DIVIDE BY 6 YEARS	29750.00 (D)
DIVIDE BY 72 MONTHS (E)	413.19

OPERATING COSTS

(11) FUEL	305.10
OIL	(12) 10.57
(13) LUBE	10.57
TIRES	(14) 294.14
(15) PARTS	1504.95
LABOR	(16) 21.47
(17) FINES AND CITATIONS	6.07
TRAILER	(18) NONE
(19) OTHER	0
SUB-TOTAL	(20) $2152.85

COSTS

YEARLY	(27) $72900.48
(28) MONTHLY	6075.04
WEEKLY	(29) 1412.80
(30) DAILY	282.56
HOURLY	(31) 35.32

AVERAGES (32)

20 GAL. DIESEL (DAY)
1 QT. OIL (DAY)
1 TUBE LUBE (DAY)

LABOR COSTS

(21) SALARY	1890.00
INSURANCE	(22) 577.80
(23) LABOR COSTS	708.75
OTHER	(24) 0
(25) SUB-TOTAL	$3176.55
TOTAL	(26) $6075.04

4. Interest on payment (if known and easily separated from No. 3).
5. Depreciation. (Use this item only if no monthly loan payment is being listed.) If you decide to use depreciation instead of monthly installment payments, items *a* through *e* give you one method of figuring depreciation.
   a. Present market value of the equipment.
   b. Deduct a 15% salvage value.
   c. The answer is the depreciable value.
   d. Divide *c* by 6 years (the expected life span of the equipment).
   e. Divide *d* by 72 months (number of months in 6 years). You now have the monthly amount that can be used for depreciation.
6. Property tax. (Take annual figure and divide by 12 months.)

7. License plates. (Divide annual fee by 12 months.)
8. Insurance. (Divide annual fee by 12 months.)
9. Radio. (Cost of unit and system charge annually divided by 12 months.)
10. Subtotal of Nos. 3 through 9.
11. Fuel costs per month.
12. Oil cost per month.
13. Lubricant costs per month.

COST OF OPERATING 19 80

	Initials	Date
Prepared by	[illegible]	3/1/80
Approved by		

★ DIVIDED BY NUMBER OF SIDES

ITEM		MONTHLY	WEEKLY	DAILY	HOURLY
		21.3	4.3	8	
Pick-up Truck		1214.10	282.35	57.00	7.13
Mobil Radios	★	1083.33	251.94	50.87	6.36
Accounting/Legal	★	1100.00	255.81	51.64	6.46
BU-80 Tower		7728.49	1797.32	362.84	45.36
BU-90 Tower		7728.49	1797.32	362.84	45.36
Bantam 266		6075.04	1412.80	282.56	35.32
Marion Shovel		5325.00	1238.37	250.00	31.25
Linkbelt LB-98		5772.30	1342.40	271.00	33.88
FMC		5410.20	1258.17	254.00	31.75
AC-16		5100.00	1186.05	239.44	29.93
HD-21		6390.00	1486.05	300.00	37.50
HD-955		5857.50	1362.21	275.00	34.38
D-8 Cat		6177.00	1436.51	290.00	36.25
Ranger 667		4153.50	965.93	195.00	24.38
AC-Grader		5100.00	1186.05	239.44	29.93
Lowboy		2268.45	527.54	106.50	13.31
Tree Puller		1810.50	421.05	85.00	10.63
Booster 'PUP'		532.50	123.84	25.00	3.13
Fire Truck		2130.00	495.34	100.00	12.50
Crew Bus		1214.10	282.35	57.00	7.13
Rigging	★	7000.00	1627.90	328.64	41.08
Equip. Insurance	★	2815.80	654.84	132.19	16.52
Personnel Insurance	★	12812.00	2979.53	601.50	75.19
Fire Trails/Slash	★	6500.00	1511.63	305.16	38.15
Maint. & Repairs	★	16283.00	3786.74	764.46	95.56
Shop & Office	★	2000.00	465.12	93.90	11.74
Fallers/Buckers		SUBCONTRACTORS ACTUAL COST			
Miscellaneous	★	1000.00	232.56	46.95	5.87
Labor Cost	★	50000.00	11627.90	2347.41	293.42
Fed.,State & FICA	★	24163.35	5619.38	1134.43	141.80
Relocation Cost	★	400.00	93.02	18.79	2.35
PUC & Permits	★	2000.00	465.12	93.90	11.74
Citations & Fines	★	97.84	22.75	4.60	.58
Log Truck, New		5628.06	1308.85	264.23	33.03
Log Truck, Old		4423.34	1028.68	207.67	25.96
Log Hauling (GYPO)	★	61000.00	14186.05	2863.84	357.98
Safety Equipment	★	1350.00	313.95	62.79	7.85
Saws & Repairs	★	700.00	162.79	32.86	4.11

Figure 28. Cost of operating list

14. Tire costs (if machine is rubber tired).
15. Parts costs (if any repairs were necessary on old equipment).
16. Labor cost for repairs and maintenance.
17. Fines and citations. (Divide annual cost for individual vehicle by 12 months.)
18. Trailer costs.
19. Miscellaneous costs.
20. Subtotal of operating costs.
21. Monthly salary of operator (base salary only, no fringe).
22. Cost of insurance for operator (worker's compensation only).
23. Cost of operator withholdings.
24. Miscellaneous costs for operator.
25. Subtotal of labor costs.
26. Total of Nos. 10, 20 and 25.
27. Item 26 total times 12.
28. Item 26 total.
29. Item 28 divided by 4.3 weeks per month.
30. Item 29 divided by 5 days per week.
31. Item 30 divided by 8 hours per day.
32. Averages for fuel and lubricants (taken from Equipment Use Form).

After completing one of these forms for each piece of equipment, put the final results (Nos. 28 through 31) on the Cost of Operating List, as shown in Figure 28. The Cost of Operating List is the second thing that makes our input complete.

Now we are ready to complete Form No. 16, the Project Cost Analysis Report.

Looking at this report we can now realize the importance of all the detail we have gathered in each of the timber harvesting elements. Now we can accomplish a step-by-step coverage of how to prepare the detailed cost analysis.

Each entry on Form No. 16A is accomplished in the following manner:

1. Job number assigned at the beginning of the project from master control sheet.
2. Date the analysis is prepared.
3. Contracting agency or mill.
4. Geographic location of job (sale name and/or map coordinates).
5. Date project was started.
6. Date project was finished (when applicable).
7. Brand being used to identify the logs.
8. Period covered by the report.
9. Number of working days for this report.
10. Log volume in board feet for the period of the report. This figure can be obtained from the scaling bureau reports for the appropriate period. If no reports are available for the period in question, you can extend a dozen or so daily load tickets to determine a load average. Then multiply the average times the number of loads each day for the period of the report. This information is obtained from Form No. 9, the Daily Load Report. This exercise will give you a very close approximation of the actual volume for the period.

Form No. 16. Project cost analysis report

PROJECT COST ANALYSIS REPORT

JOB NUMBER *12* (1)

(2) DATE *5/19/81* CONTRACT FOR (3) *AMAZON WOOD PRODUCTS*

JOB LOCATION (4) *COUGAR CREEK* START *3/15/81* (5) FINISH (6)

(7) BRAND *A4* PERIOD OF REPORT *MAY 5-17, 1981* (8) WORKING DAYS *10* (9)

FOOTAGE FOR PERIOD *1,120,000* (10)

EQUIPMENT ANALYSIS	NO.	CONTROL NO.	COST/PERIOD	COST/DAY	COST/M	NO. DAYS	
(11) LOADERS	1	L598	2292.80	229.28	2.04	10	
(12) SKIDDERS	(A)	(B)	(C)	(D)	(E)	(F)	
(13) TOWER	1	BU-80	3057.10	305.71	2.73	10	
(14) GRADER	1	HD-16 p/s	480.00	160.00	.43	3	
(15) FIRETRUCK	1	MACK	300.00	30.00	.27	10	
(16) PICK UPS	4		2280.00	228.00	2.04	10	
(17) CREW BUS	1		400.00	40.00	.36	10	
(18) MISC. TRK.							
(19) MISC. EQP.							
(20) SAWS	3	HOMELITE	630.00	63.00	.56	10	
(21) RIGGING			135.00	13.50	.12	10	
(22) RADIOS	6		720.00	72.00	.64	10	
(23) SAFETY			209.30	20.93	.19	10	
(24) LOG TRKS.	5	TM	10383.50	1038.35	9.27	10	
(25) LOG TRKS.	2	GYPO	9000.00	900.00	8.04	10	60 LOADS
(26) EQUIPMENT TOTAL			29887.70	3100.77	26.69		

OTHER EXPENSES	COST/PERIOD	COST/DAY	COST/M	NO. DAYS
(27) ACCOUNTING	360.00	36.00	.32	10
(28) FEES, TAGS, PERMITS	45.30	4.53	.04	10
(29) INSURANCE-EQUIPMENT	305.00	30.50	.27	10
(30) INSURANCE-PERSONNEL	200.00	20.00	.18	10
(31) EQUIPMENT DOWN-TIME	36.30	3.63	.03	10
(32) RELOCATION COST	56.30	5.63	.05	10
(33) SUB-TOTAL	1002.90	100.29	.89	

11. Tractors or loaders being used:
    a. Number being used.
    b. Type being used.
    c. Total cost of equipment for period (using the cost per day figure from the Cost of Operating List shown in Figure 28).
    d. Cost per day from Cost of Operating List (Figure 28).

Form No. 16. *Continued*

PROJECT COST ANALYSIS REPORT (2) JOB NUMBER (34) 12 DATE 5/19/81 (35)

OTHER EXPENSES, CONTINUED	PERIOD	DAY	PER-M	NO. DAYS
(36) FIRE TRAILS, SLASH, ETC.	300.00	30.00	.26	10
(37) LABOR (COMPANY)	7509.50	750.95	6.70	10
(38) LABOR (SUBCONTRACTOR)	(C)	(D)	(E)	(F)
(39) PAYROLL COSTS (FICA, ETC.)	2816.00	281.60	2.51	10
(40) FUEL & LUBE OIL	100.00	10.00	.09	10
(41) SUPERVISORY	2350.00	235.00	2.10	10
(42) MAINTENANCE/REPAIRS	100.00	10.00	.09	10
(43) SHOP EXPENSE 25 %	500.00	50.00	.45	10
(44) CUTTERS (COMPANY)				
(45) CUTTERS (SUBCONTRACTOR)	3750.40	375.04	3.35	10
(46) MISCELLANEOUS	55.00	5.50	.05	10
(47) PAGE ONE SUB-TOTAL	1002.90	100.29	.89	
(48) PAGE ONE & TWO SUB-TOTAL	18483.80	1848.38	16.49	
(49) EQUIPMENT TOTAL	29887.70	3100.77	26.69	
(50) GRAND TOTAL	48371.50	4949.15	43.18	
OTHER INFORMATION				
(51) FOOTAGE	1120000	112000		
(52) EARNINGS	72800.00	7280.00	65.00	
(53) HOLDBACK	1120.00	112.00	1.00	
(54) GROSS	71680	7168.00	64.00	
(55) MINUS TOTAL EXPENSES	48371.50	4949.15	43.18	
(56) PROFIT OR (LOSS) PERIOD	23308.50	2218.85	20.82	
NOTE (57) GROSS MERCH.	X	=		
(58) NET UTILITY	X	=		
(59) CULL OUTS	X	=		
(60) UTILITY DIFF. GROSS/NET	X	=		

e. Cost per thousand board feet. (Divide the cost for the period by the footage for the period: $2,292.80 divided by 1,120,000 equals $2.04 per thousand board feet.)

12 through 20 are completed in the same manner as just described for items 1 through 11.

21. Rigging costs can be computed in two different ways:
   a. The actual job cost can be determined by reviewing the rigging control sheets (Form No. 8) for the period in question and comparing the entries with the appropriate invoices to determine the cost.
   b. Take the total rigging cost for the period and divide by the number of sides active during the period. This will average the cost out to each job going on at that time.
22. Radio costs are taken from the Cost of Operating List, as in items 11 through 20.
23. Safety programs and equipment. The overall cost of safety programs and equipment should be divided equally among job sides. In our example, the daily cost of safety from the Cost of Operating List was divided among three separate job sides.
24. Log trucks (your own). The number of log trucks can be determined by consulting your Daily Load Reports (Form No. 9). The Cost of Operating List will contain your daily costs for one truck. Simply multiply the number of trucks used times the daily cost for one truck. You will notice in Form No. 15 that a financed log truck will obviously cost more to operate than one that is not financed.
25. Log trucks (subcontractors). As in item 24 the number of trucks can be determined by using the Daily Load Reports (Form No. 9). The cost is determined by your method of payment to the subcontractor. If you pay by the load, it is simple. If you pay by the thousand board feet, it is more involved. By the thousand will require checking the bureau tickets and figuring the volume for each load hauled by a subcontractor.
26. The totals of items 11 through 25.
27. Accounting costs. The total cost for the period should be divided equally among job sides.
28. Fees, tags and permits (costs specifically related to this project only).
29. Any insurance costs not figured into the Cost of Operating Analysis forms on each piece of equipment.
30. Any personnel insurance not included under operator costs on the Cost of Operating Analysis forms. This would include choker setters, rigging slingers, chasers and any other persons not involved in equipment operation.
31. Equipment downtime is an expense that should be averaged and divided equally among job sides, unless you wish to track down specific charges on specific pieces of equipment. If you wish to do this, simply check the Equipment Use forms for the equipment that was used on this job during this specific period of time.
32. Relocation costs can be obtained from the Job Relocation and Equipment Move form shown earlier (Form No. 2).
33. Subtotal of other expenses, items 27 through 32.

Now you will be using Form No. 16B for page two of the report.

34. Job number.
35. Date of report.

36. Cost of fire trails and cleanup. This cost is usually not figured until the final analysis is accomplished. This is usually the last thing done on any logging job.
37. Labor costs (your own). This information can be taken from the Manpower Utilization Summary (Form No. 7). This will reflect only the base salary of the crew. Remember to exclude the equipment operators' salaries.
38. Labor costs (subcontractors). This information can be furnished by your subcontractor. It will usually be an analysis by the thousand board feet or by the hour.
39. Any withholdings not included in item 37 above.
40. Fuel and lube oil (only costs associated with heavy equipment not included in the individual breakdowns in items 11 and 25).
41. Cost of supervisory personnel (monthly or salaried) associated with completing the job.
42. Cost of maintenance and repairs not included in individual breakdowns, items 11 and 25.
43. Daily shop overhead expense divided by number of job sides.
44. Cutter payroll (your own). If you do your own cutting, the costs of the people can be taken from the Manpower Utilization Summary (Form No. 7).
45. Subcontract cutters' payroll (if you wish to separate it from item 38 above).
46. Any items not included in any other major category.
47. Risk factor or other random variables you wish to include.
48. Subtotal of all columns from pages 1 and 2.
49. Total from page 1.
50. Grand total of all columns pages 1 and 2.
51. Gross footage from total period of production this report.
52. Total earnings for period covered by this report (taken from cash receipts journal).
53. Total amount held by agency or mill (per thousand) for volume delivered.
54. Total gross income for period of report.
55. Subtract the total for expenses.
56. The net profit or (loss) for the report period.
57. Gross Merch. total footage (from bureau reports).
58. Net Utility total footage (from bureau reports).
59. Total footage, cull outs (from bureau reports).
60. Utility differential gross/net (from bureau reports).

After these reports are completed and evaluated, they should be filed in the job folder along with all other records of each project.

After the total cost analysis report has been completed, the data can be converted to a visual form. This will help you to visually see the comparison of various costs to each other. It will also be a valuable aid in explaining your job performances to others (such as the bank) who might be critically interested.

To accomplish this end we developed Form No. 17, the Cost Distribution Analysis. This form shows both statistical data and visual data. It is helpful many times in determining where your attention should be focused in your cost reduction efforts.

Form No. 17. Cost distribution analysis

COST DISTRIBUTION ANALYSIS

DATE *6/1/81* (1)

TOTAL AMOUNT SPENT THIS PERIOD *19 922.15* (2) BRAND *TW4* (3)

CONTRACT FOR: (4) *TWILIGHT VENEER CORP.*

PERIOD OF REPORT (5) *MAY 1-12, 1981* WORKING DAYS *8* (6)

JOB NO. *38* (7)

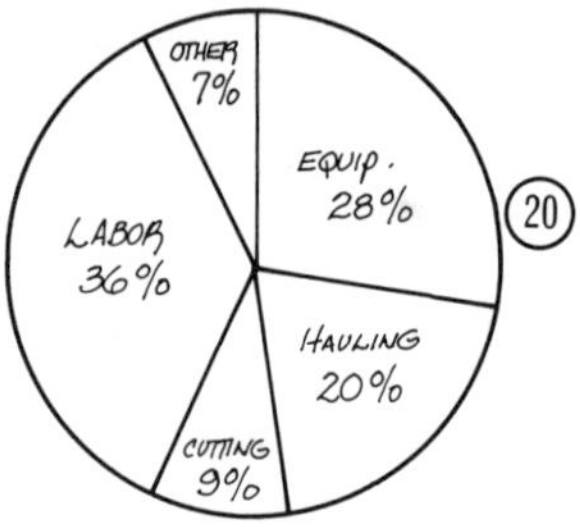

(8) EQUIPMENT COST	*5632.31*	(9)	*28* %
(10) HAULING COST	*3948.92*	(11)	*20* %
(12) CUTTING COST	*1873.44*	(13)	*9* %
(14) LABOR COST	*7223.23*	(15)	*36* %
(16) OTHER COST	*1244.25*	(17)	*7* %
(18) TOTAL COST	*$19 922.15*		100%

NOTES: (19) *SINCE LABOR WAS THE PRIMARY EXPENSE DURING THIS PERIOD A REVIEW SHOULD BE MADE OF OVERTIME & THE NUMBER OF PERSONNEL ASSIGNED, PERHAPS WE COULD LOWER THE LABOR EXPENSE & THEREBY INCREASE OUR PROFIT ON THIS PROJECT.*

The form is simple to use and should be filled in in the following way:

1. Date of report.
2. Total amount of money spent during this period.
3. Brand being used to identify the logs.

4. Agency or mill you are working for.
5. Period covered by the report.
6. Number of working days during the report period.
7. Job number assigned from master file.
8. Total cost for equipment expenses.
9. Percent of total expenses.
10. Total cost for hauling expenses.
11. Percent of total expenses.
12. Total cost for cutting expenses.
13. Percent of total expenses.
14. Total cost for labor.
15. Percent of total expenses.
16. Other (miscellaneous) cost total.
17. Percent of total expenses.
18. Same as line 2.
19. Notes relevant to report analysis.
20. Plotting of graph using the percentage protractor shown in Figure 29. This protractor can be used to plot any pie chart. The percentages are marked off in 5's. By turning the half circle around, you get the full 100% circle.

When completed, this form can be filed along with all the other data in the project folder.

Element 13: Unloading

Element 13 is usually handled where the logs are delivered, at no expense to the contract logger. If, however, you have to figure this cost, you would use the various forms regarding equipment and personnel that we have already described in other elements. Specifically, you would use Form Nos. 4, 5, 6, 7 and 10.

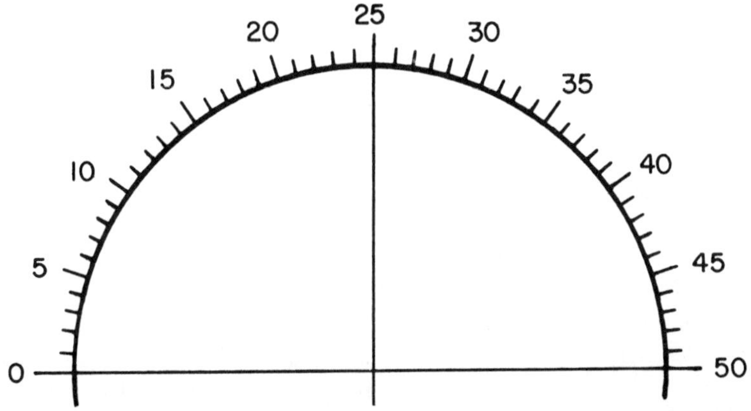

Figure 29. Percentage protractor

Elements 14 and 15: Cleanup, Slash Disposal and Reforestation

If you accomplish Elements 14 and 15 using your own personnel, you can determine the cost by using Form Nos. 4, 5, 6 and 7 for the period of days involved in the accomplishment of the element.

If this work is subcontracted, you would require your contractor to provide a total cost breakdown for your files.

The cost per thousand board feet is of interest to you since this cost should be included in your bid price for logging each job. As you may be aware, this is a substantial cost and must be included in your logging price. Some mills take care of this element; if so, it is better for you.

Element 16: Final Inspection

Element 16 is usually handled by the agency or mill. After the final inspection has been made and the unit approved for release, you will usually receive a check for all the money held back during the logging contract period. Not all mills will hold back funds to ensure your integrity for the cleanup, but it seems most require at least $1.00 per thousand board feet.

It would be good if you or one of your people escorted the inspector during this final phase. Many times, problems can be resolved as the inspection is being made. Often, additional work is required to obtain the release. This work can be held to a minimum by your representative being along on the inspection.

Element 17: Equipment Removal, Transport, and Relocation

You will follow the same procedure in removing the equipment that you used in getting it to the job site. Refer to the discussion of Element 5. The same forms are used.

Element 18: Closing out the Job Folder

During the course of accomplishing the job, you have been putting all records related to it in the job folder. At the end of each job, all such materials should remain a part of the permanent file for each job.

When all work is finally completed and all related matters are cared for, you will move the file folder into the "jobs completed" section. It will remain there as long as you are legally required to keep the records or as long as you wish to have the information available to you.

Element 19: Final Financial Analysis

Under the procedure outlined in Element 12, fill out a Form No. 16.

Do an analysis for the entire time period covered by the project. This will give you totals on everything related to the job and a final look at your overall financial performance. If you did everything right, you will have a profit rather than a loss.

If there was a loss, you will want to review the entire project folder with your bookkeeper and your accountant. By doing this you can find the mistakes you have made with either personnel or equipment and make sure you don't make the same mistakes again.

MANUAL OR MACHINE?

The procedure outlined here is a manual procedure. The entire exercise can also be done by a computer. You will still need the forms described in the various elements to get the information from the woods to the office. So, the forms in the system provide the input and a coding system can be developed to program the entire cost analysis exercise into the computer.

If you feel it is worthwhile, you should look into the use of computers for your logging business. Some loggers buy their own small computers; others hire a computer service. It is really up to you as the boss to determine how you want your information prepared. If you are very small, the manual method is by far the most economical. If you are growing rapidly, the computer service can help you to continue to grow with fewer worries along the way.

SUMMARY

It is our feeling that this system can be tailored to any logging operation, large or small.

The extent of involvement is in your hands. The more data you have, the deeper and more accurate the analysis.

If the employees do not support the system and fill out the forms on a daily basis, the system will fail. You must have the daily inputs to make it work. Once employees establish the habit of filling out the forms, they rarely forget to complete them or to turn them in. It is a matter of discipline and a matter of your setting the proper example. Make the completion of the forms a condition of employment. If they don't fill out the forms, they don't work. Believe me, they will fill out the forms, and it will work.

Remember: *strength through discipline.*

CHAPTER 9

The People Part of Your Business

Years ago my father told me, "Ronnie, there's nothing wrong with this old world. It's the darned people in it that foul everything up." Throughout life I have found that to be the case. People can be very frustrating.

Indeed, people can be the most frustrating part of our lives and our business. But, as we all know, we can't get a thing done without them. I have yet to see a one-man logging show. Since we have to deal with them and keep them happy, there are certain things we must take into consideration.

We are told that a good manager is one who can get things done through other people. Before you can get people to work, you have to find them, recruit them and then keep them happy while they work for you. This chapter offers some suggestions and proven ideas that will help you to find, maintain and keep good employees for your company.

COMPANY POLICY MANUALS AND EMPLOYEE HANDBOOKS

For an employee to be happy, he has to know what you expect of him and how he is to conduct himself during his time with your company.

Where feasible, a company should have an employee handbook or a company policy manual that spells out every policy concerning the actions of your personnel.

These valuable working tools not only provide instruction and information to the employee, but they also provide a great deal of protection for the company and its management. We find that where policy is not clear, there are serious problems with labor disputes, discrimination charges and other serious misunderstandings between labor and management. The small investment a company makes in such handbooks and manuals is far offset by the benefits they produce in better employee-employer relationships.

Let us now explore some of the most common policies regarding the people part of your business.

FINDING AND HIRING GOOD EMPLOYEES

Perhaps two of the most difficult areas of personnel management are finding and hiring good people. How do you get the best person for the job? Once you have the best person, how do you keep him happy? Years ago, things were very simple. You hired only the people you wanted to hire and you fired the people who couldn't get the job done. Not so today!

Things today are not quite as simple as they were in our fathers' time. Today we have equal opportunity laws, fair labor practice laws, worker's compensation laws, safety and training laws and a multitude of other laws that we as business managers have to contend with. To keep people happy today, we must know about these laws and make sure we are in compliance with them.

So, before we get into the policies a small logging company should set up, let's briefly review some of the laws that we must obey in the process of establishing those policies.

Affirmative Action and Equal Employment Opportunity

Henry Becque once said, "The defect of equality is that we only desire it with our superiors." George Orwell said, "All animals are equal, but some animals are more equal than others." These two quotations help us to understand that true equality is a paradox.

This paradox has prompted government to give us the Affirmative Action and Equal Employment Opportunity laws. Concerning this subject, U.S. Supreme Court Justice Warren Burger made the following comment: "What is required by Congress is the removal of artificial, arbitrary and unnecessary barriers to employment when the barriers operate hatefully or unjustly to discriminate on the basis of racial or other impermissible classification. Title VII of the Civil Rights Act proscribes not only overt discrimination but also practices that are fair in form but discriminatory in operation."

For those of us in the logging business, the first thing we must do is to make a necessary distinction between Affirmative Action and Equal Employment Opportunity. Most of us are aware that Equal Employment Opportunity is the law, both federal and state, in the United States. For example, the Equal Employment Opportunity Act of 1964 and the strengthening amendments added in 1972 provide federal legal enforcement.

Title VII of the act prohibits discrimination because of race, color, religion, sex or national origin in all employment practices, including hiring, firing, promotion, compensation and other terms, privileges and conditions of employment.

As amended, Title VII now covers:

- All private employers of 15 or more people.
- All educational institutions, public and private.
- State and local governments.
- Public and private employment agencies.
- Labor unions with 15 or more members.
- Joint labor-management committees for apprenticeship and training.

Most countries in the world have similar laws to protect their working people.

But what about Affirmative Action and written Affirmative Action Plans? Who must have them and who does not need them? In the United States, those in business who have a federal timber contract or subcontract are naturally interested in how they are affected by these laws.

The question regarding Affirmative Action Plans can be answered simply with what is called the 50/50 requirement. If you have a government contract or subcontract for $50,000 or more and you have 50 or more employees, you must have a written Affirmative Action Plan subject to government inspection.

Some might wonder about the small logger who subcontracts for the big mill with a $50,000 contract and 50 or more employees. Does the small logger have to comply with the same program that the big mill has? The answer is: *only* if the small logger has a subcontract for $50,000 or more and has 50 or more employees. If the small logger has less than 50 employees, he does not have to comply with the written Affirmative Action Program regardless of the dollar amount of his subcontract. A few years ago, the understanding was somewhat different. When talking to the local office of Federal Contract Compliance, I was told that the current administration was conducting an extensive review of the 50/50 requirement with the intention of raising that requirement to higher figures. If this happens, it would relieve many more companies of the financial burden of administering written Affirmative Action Plans.

For those who must comply with the Affirmative Action Plan, here is a condensed version of what you must do:

- An executive within your company must be appointed as Director or Manager of the Equal Opportunity Program.
- A policy statement must be written expressing the company's attitude on the subject, assigning overall responsibility and providing for a reporting and monitoring procedure.
- The policy statement must become a part of the company policy manual.
- The policy must be publicized in the company newspaper, magazine, annual report, newsletter and other applicable media.
- The company must conduct special meetings with all executive, management and supervisory personnel to explain the intent of the policy.
- The company must schedule special meetings with all employees to explain the overall policy.
- The company must inform all recruiting sources verbally and in writing of the policy.
- The company must notify minority and women's organizations, community agencies, community leaders, secondary schools and colleges of the company policy—in writing.
- The company must display posters on all company bulletin boards regarding the Affirmative Action Program.
- The company must survey and analyze minority and female employment by department and job classification.
- The company must identify areas of underutilization and concentration.
- The company must determine the extent of underutilization of minorities and females.
- The company must develop goals and timetables.
- The company must identify the causes of underutilization if such causes exist.
- The company must develop and implement specific programs to eliminate discrimination barriers and reach the goals established.

- The company must establish an internal audit and reporting system to monitor and evaluate the progress in each aspect of the program.

It is quite obvious that compliance with this program is going to be expensive and time consuming. But those who meet the 50/50 requirement must comply with the many details that accompany the law. The overall law is much more complicated than this brief list. If you must comply, obtain a copy of the entire law and make sure that your company complies with every detail.

Sample Policy Statement

April 4, 1982

Policy statement from the office of the ABC Company.

It is our intention to bring the ABC Company into full compliance with the goals and objectives of the Federal Equal Opportunity Employment rules and regulations.

The first of several steps involved is to appoint John Doe as Director of the ABC Affirmative Action-Equal Opportunity Program. Mr. Doe will be responsible for implementing our effort to eliminate the possibility of anyone being discriminated against because of race, color, religion, sex or national origin. Our total program is designed to ensure equality of opportunity in all aspects of company employment. Affirmative Action will be taken by Mr. Doe to make the appropriate agencies aware of our decision in this matter.

This policy statement will become a permanent part of ABC policy and the charter for our formal program. A quarterly report will be prepared regarding the effectiveness of our program and this report will be used as a tool to monitor our continued progress.

Signed by the Head of ABC Company

Fair Labor Practices and Wage and Hour Laws

Regulations for your state that deal with labor law can be obtained by calling or writing your state capital. The complete content of these laws is far beyond the scope of this book. But you must be very careful in all your dealings with your employees, both monthly and hourly. Remember that complete documentation is a must in areas of hiring and firing, time worked, compensation, seniority, layoffs, bonuses and raises. Complete personnel files will be very valuable to you in all cases of this nature. Every dollar you spend in record keeping will be worth its weight in gold if you ever have a fair labor dispute filed by an employee.

Worker's Compensation Laws

Worker's compensation insurance is a requirement in every state. Each state puts out a manual for employers. It is vital that you obtain a copy of this manual for your company files. All requirements are clearly outlined for you. The manual also contains other information vital to the operation of your company.

Safety and Training Laws

You must obtain copies of your state's program for occupational safety and health as it applies to logging. There will usually be a specific manual prepared for loggers and you will be expected to comply with all the rules and regulations listed. Many associations have programs already designed to comply with your state's laws. Some compensation insurance carriers have plans available and approved by your state. Make sure you obtain the manuals and the programs and comply with them. The fines for noncompliance are usually very severe, even for first offenders.

Now that we have covered some of the basic laws you must contend with, let's return to our discussion about finding and hiring good people.

Advertising for New Employees

When advertising for a new employee, make sure to give as much information as possible about the job experience required to perform the job in a satisfactory manner. Outline the specific job and what it entails. It is also good to list a salary range and to mention that the salary is governed by the experience of the applicant. It certainly would be to your advantage to mention that you are an equal opportunity employer.

If you have mandatory compliance and an Affirmative Action Plan, make sure to save copies of the ads for your affirmative action file.

Applications and Interviews

Always have every prospective employee fill out an application and a health questionnaire. This is vitally important to you and your company. As we have already mentioned, careful documentation is worth its weight in gold.

Form No. 18 is an example of a good application. It contains everything you need to know about an employee. If you don't already have an application form on all of your employees, get them all to fill out a current application.

If you already have applications on your current employees, it may be time for an update of their personnel records. This should be done periodically. You can use Form No. 19, Personnel Record Update. These forms should be done at least once a year for each employee.

A Health Questionnaire (Form No. 20) should be filled out at the same time the application is completed by the prospective employee. This form will be valuable in determining insurance eligibility requirements and certain worker's compensation factors. Make sure that everyone has one of these in his personnel folder. Many times the history of the person will reflect a negative pattern, such as chronic back injury. Knowing these things can be a protection to you against fraudulent insurance claims.

The interview is an important time for you and the employee. This is the time that you both decide if you want to make a committment. Form No. 21 is a form that can be used for the employment interview. This form is used only as a guide to determine as much as you can about the prospective employee and his background. If you have a staff member who interviews prospective employees, you may wish to have him prepare an Interview Report Form (Form No. 22). This report is made up of the staff member's observations and allows you to make the decision from his observations. If the prospective employee is hired, all the forms would become a part of his personnel file.

Form No. 18. Employment application form

EMPLOYMENT APPLICATION FORM
(We are an equal opportunity employer)

Date 4/21/82

Full Name RONALD P. WEBER Date of Birth 11/4/47

Street Address 8608 N. 31ST. DRIVE City SPRINGFIELD

County LANE State OREGON Zip 97477

Telephone where you may be contacted: 746-8995

Social Security No. 512-30-7357 Driver's License No. A037585

State where issued OREGON Expiration date NOV. 84

Job you are applying for: HOOKTENDER

Years Experience 15 Other positions you are qualified for

LOADER OPERATOR, GRADER & CAT

Who referred you? LLOYD WILSON Rate of pay 10.50

I am applying for: [X] Full time work [ ] Part time work

I can start on: 5/1/82 I am willing to relocate [X] Yes [ ] No

In case of an emergency, please notify CARMELLA WEBER

Relationship WIFE Address SAME AS ABOVE

Telephone SAME

I am a U.S.Citizen [X] Yes [ ] No I have served in the Military [X] Yes [ ] No

I am currently in Guard or Reserves [ ] Yes [X] No

Please circle highest grade completed in school: 1 2 3 4 5 6 7 8

9 10 11 (12) Name and location of High School LOWELL HIGH SCHOOL, LOWELL, OREGON

Please enter below any schools, degrees, training and/or experience you may have that is related to the job you are seeking:

SPECIAL COURSE IN EQUIPMENT OPERATOR CERTIFICATION
LANE COMMUNITY COLLEGE

Form No. 18. *Continued*

Page two. EMPLOYMENT APPLICATION FORM

This information is needed for health and accident insurance:

Married ☒ Single ☐ Living Together ☐ Divorced ☐ Separated ☐

Is spouse or companion employed? Yes ☐ No ☒

If employed list Company Name ____________

First Name of spouse or companion CARMELLA Date of Birth 5/16/49

Number of dependents 3 Ages of dependents 10 & 5

AFFIRMATIVE ACTION INFORMATION:

This information is needed to ensure equal employment opportunity and may not be used to descriminate against you in any way.

Sex MALE Age 35 Spanish American ☐ American Indian ☐

Oriental ☐ Black ☐ White ☒ Other ☐ ____________

EMPLOYMENT HISTORY (LIST PRESENT, OR LAST EMPLOYER FIRST):

1. Employer INTERNATIONAL WOOD Address 12357 SPAR WAY, PORTLAND
   Job Title HOOK TENDER Supervisor's Name REX KEMP
   Duties (Be as Specific as possible) TENDED HOOK & HELPED WITH SIDE PREPARATION & TOWER SET UPS. SUPERVISED 5 MEN.
   Reason for Leaving WANT TO WORK FOR A SMALLER COMPANY
   EMPLOYED: From 2/77 To PRESENT May we contact? Yes ☒ No ☐

2. Employer FARAWAY FIRS CO. Address 98765 FOX HOLLOW RD., DEXTER
   Job Title GRADER OPERATOR Supervisor's Name HAROLD COLE
   Duties (Be as Specific as possible) OPERATED GRADER ON ROAD BUILDING PROJECTS, ALSO OPERATED HD-16 CAT
   Reason for Leaving FOUND A BETTER JOB WITH MORE RESPONSIBILITY
   EMPLOYED: From 1/67 To 2/77 May we contact? Yes ☒ No ☐

3. Employer ____________ Address ____________
   Job Title ____________ Supervisor's Name ____________
   Duties (Be as Specific as possible) ____________

Form No. 18. *Continued*

Page three. EMPLOYMENT APPLICATION FORM

Reason for Leaving ____________________

EMPLOYED: From ________ To ________ May we contact? Yes ☐ No ☐

PLEASE LIST THREE REFERENCES NOT RELATED TO YOU:

NAME	ADDRESS/PHONE	BUSINESS	YEARS KNOWN
HAROLD GRAHAM	1077 GATEWING 746-4333	INSURANCE	10
LEROY SHIELDS	815 No. OLIVE 747-7752	BANKER	18
WILLIAM KOONS	914 NO. JEFFERSON 746-1222	ART DIRECTOR	8

If you are hired, will you:

Report all injuries that you receive on the job to your supervisor?

Yes ☒ No ☐ ____________________

Abide by the safety rules of this company and the government?

Yes ☒ No ☐ ____________________

CERTIFICATE OF APPLICANT: "I hereby certify that all information in this application is as complete and correct as possible and to the best of my knowledge and belief it is truthful."

Signature of Applicant *Ronald R Weber*

Date signed: *4/21/82*

ACTION RECOMMENDED: (To be filled out by Office staff)

RECOMMEND THIS MAN FOR NEW POSITION ON COUGAR CREEK SIDE. SALARY TO START $11.60 PER HR.

[signature] 4/22/82

Form No. 19. Personnel record update form

PERSONNEL RECORD UPDATE FORM Date *3/14/82*

Name *PIO MARTINEZ* Employee # *128*

IN ORDER TO KEEP YOUR PERSONNEL RECORDS UP-TO-DATE, PLEASE SHOW BELOW ANY CHANGES SINCE *3/14/81* .(SHOW CHANGES ONLY)

Address *1357 BROADWAY APT. 27* Phone *747-7447* Marital Status:

Engaged ☐ Married ☒ Separated ☐ Divorced ☐ Widowed ☐

(Please remember that none of this information can be used to discriminate against you in any way.)

Number of dependents *0* Number of Children ____ Ages ____

Does your mate work? ☒ Yes ☐ No If yes, what kind of work *I DON'T KNOW. WE ARE SEPARATED* Their earnings: $ *?* Per ____

Have you developed any physical problems since your last update? ☒ Yes ☐ No

If yes, please describe: *MUST UNDERGO KNEE SURGERY FOLLOWING INJURY WHILE CHASING LANDING ON COUGAR CREEK*

Describe any major illnesses since last update *NONE*

PLEASE LIST ADDITIONAL SCHOOLING OR SPECIAL TRAINING YOU HAVE RECEIVED:

DATE	SCHOOL OR AGENCY	NAME OF COURSE
8/2/81	*NAT. RED CROSS*	*FIRST AID & CPR*

New memberships in technical or professional societies? *No*

New professional offices or honors *No*

Any other changes you would like us to make in your records? *No*

Employee's Signature *Pio Martinez* Reviewed by *[signature]*

Changes Made on *3/18/82* By: *[signature]*

Form No. 20. Health questionnaire form

HEALTH QUESTIONNAIRE FORM Date 4/21/82

Name RONALD P. WEBER Employee # 502

Have you had a complete physical in the last 2 years? [X] Yes [ ] No

If so, When 3/18/81 . Reason ANNUAL

Are you willing to take a physical examination now? [X] Yes [ ] No

PLEASE CHECK ANY OF THE FOLLOWING CONDITIONS YOU MAY HAVE EXPERIENCED:

Heart Problems [ ] Rheumatic Fever [ ] High Blood Pressure [ ] Varicose veins or Leg Ulcer [ ] Chest Pain [ ] Chronic Cough [ ] Tuberculosis [ ] Allergies [ ] Hay Fever or Asthma [ ] Skin Trouble [ ] Reaction to serum or drug [ ] Kidney or Bladder Trouble [ ] Ulcers [ ] Diabetes [ ] Cancer [ ] Stroke [ ] Emphysema [ ] Knee Injury [ ] Leg Pain [X] Fractures or Broken Bones [ ] Athletic Injury [ ] Hemorrhoids [ ] Shoulder Injury [ ] Hearing Trouble [ ] Ear Infection [ ] Mastoid Operation [ ] Frequent Headaches [ ] Head Injury [ ] Eye Injury [ ] Dizziness or Fainting Spells [ ] Convulsions or Epilepsy [ ] Frequent Backache [ ] Arthritis or Rheumatism [ ] Amputation of any Part of the Body [ ] Trouble with Bones or Joints [ ] Drinking Problem or Drug Addiction [ ] Nervous Condition [ ] Weakness or any body part [ ] Elbow Injury [ ] Wrist Injury [X] Severe Reaction to Bee Sting [X] Severe Poison Oak or Ivy [X]

Have you ever had a back or neck injury? [ ] Yes [X] No If so, When and Where? __________

Have you ever been rejected for medical reasons from insurance or military service? [ ] Yes [X] No If so, please explain __________

Have you ever received disability benefits from military service or the Veteran's Administration? [ ] Yes [X] No If so, please explain __________

Do you smoke? [X] Yes [ ] No Do you wear corrective lenses? [ ] Yes [X] No

I HEREBY AUTHORIZE YOU TO CONSULT ANY PREVIOUS EMPLOYER OR DOCTOR WHO HAS TREATED OR EXAMINED ME. [X] Yes [ ] No

CERTIFICATE OF APPLICANT: "My signature below certifies that all information in this application is complete and correct to the best of my knowledge and belief."
SIGNATURE OF APPLICANT Ronald P. Weber

Form No. 21. Employment interview guide form

EMPLOYMENT INTERVIEW GUIDE FORM Date 4/22/82

Name of Applicant RONALD P. WEBER

Being Considered for HOOKTENDER
Job Title

This interview guide has been prepared to help in the selection of employees by making each interview more comprehensive. Remember this form is just a guide to help direct questions to the applicant that will reveal the information needed to determine whether the prospective employee will be the right person for the job. You may skip some of the questions or add questions as you deem necessary during the actual interview process.

WORK HISTORY (Circle yes or no)

1. Please describe the kind of work you did in your last job. (You will be interested primarily in the last five years.)

   APPLICANT HAS EXTENSIVE EXPERIENCE AS A HOOKTENDER. KNOWS HOW TO LEAD MEN & DESIGN LANDINGS. HAS A VERY GOOD ATTITUDE & HAS A STEADY EMPLOYMENT RECORD.

2. May we contact your former employers for references? (Yes) No
3. Has any former employer refused to give you a recommendation? Yes (No)
4. What experience have you had in handling people? SUPERVISED 5 PEOPLE WITH INTERNATIONAL WOOD
5. What did you like best in your last job? DEALING WITH FELLOW EMPLOYEES
6. What did you dislike about your last job? PAY WAS TOO LOW & COMPANY TOO BIG
7. Why did you leave your last job or why are you leaving your present job? I NEED A CHANGE
8. Have you ever been criticized by your former employers? Yes (No)
9. Do you like the kind of work you have chosen to do? (Yes) No

Form No. 21. *Continued*

Page two. EMPLOYMENT INTERVIEW GUIDE FORM.

PERSONAL INFORMATION (Circle yes or no)

10. Have you ever had a part-time job to supplement your income? (Yes) No

11. Do you own your own home? (Yes) No

12. What are your goals for the future? TO CONTINUE MY EDUCATION & IMPROVE MYSELF IN THE AREA OF SUPERVISION.

13. Are you satisfied with your progress in your chosen field? (Yes) No

HEALTH INFORMATION (Circle yes or no)

14. What illnesses, accidents or operations have you had in the past two years?

NONE

15. Do you have any physical defects that would hinder you in doing this kind of work?

NO

16. Have you ever received compensation for injuries? Yes (No)
(If yes, please describe)

17. What is the general state of your health at this time?

EXCELLENT

EDUCATION INFORMATION (Circle yes or no)

18. Have you continued your education in any way since leaving school?

(Yes) No If Yes, How? NIGHT CLASSES IN SUPERVISION AT OREGON STATE

19. Have you had any specialized training in connection with your job?

(Yes) No If Yes, What? TRAINING IN SIDE PREPARATION & EQUIPMENT SET UP WITH LOGGING ASSOCIATION

20. Do you have plans to continue your education in the future?

Form No. 21. *Continued*

Page three. EMPLOYMENT INTERVIEW GUIDE FORM.

(Yes) No If Yes, How? ENROLLMENT AT LCC FOR NIGHT CLASSES

OTHER INFORMATION (Circle yes or no)

21. Do you belong to any organizations or clubs? Yes (No) If yes, please list.

22. What do you consider to be your strong points? THE ABILITY TO GET THINGS DONE THROUGH OTHER PEOPLE & A STRONG SAFETY RECORD

23. What are your hobbies? HANDBALL & FISHING

24. What is your feeling about our company? I LIKE THE FACT THAT IT IS SMALL. YOUR EQUIPMENT IS IN EXCELLENT SHAPE.

25. What do you expect to gain by working for our company? LEARNING MORE ABOUT SLACKLINE LOGGING & THE OPERATION OF A PRODUCTIVE SIDE.

You may insert any other questions you feel relevant:

[signature]
Interviewer

Form No. 22. Prospective employee interview report form

PROSPECTIVE EMPLOYEE INTERVIEW REPORT FORM Date 4/22/82

Name of Applicant RONALD P. WEBER Telephone 746-8995

Address 8608 N. 31ST DRIVE

Candidate for: HOOKTENDER Interview: (1) 2 3

Person doing interview RRM

AS YOU CONDUCT THE INTERVIEW CHECK YOUR IMPRESSIONS IN THE MOST APPROPRIATE BOX IN EACH AREA:

EXPERIENCE	No relationship between background and job ☐	Fair background ☐	Average background and experience ☐	Background considerable; very good experience ☐	Excellent experience and background ☒
FRIENDLINESS	very distant and aloof ☐	Approachable; friendly ☐	Warm; friendly ☐	Very outgoing ☒	Extremely friendly ☐
POISE	appears nervous and Ill at ease ☐	Somewhat tense; is easily irritated. ☐	average poise ☐	Sure of himself ☐	Extremely well composed ☒
PERSONALITY	Unsatisfactory ☐	Questionable ☐	Satisfactory ☐	Very desirable ☒	Outstanding ☐
CONVERSATIONAL ABILITY	Expresses himself poorly. ☐	Tries to express himself ☐	Average fluency ☐	Talks well ☐	Excellent expression ☒
ALERTNESS	Slow ☐	Rather slow ☐	Grasps with average ability. ☐	Quick to understand ☐	Exceptional ☒
GENERAL WORK FIELD	Poor knowledge ☐	Fair knowledge ☐	the average applicant ☐	Fairly well informed ☐	Has excellent knowledge ☒
APPEARANCE	Very untidy ☐	Somewhat careless ☐	Satisfactory ☐	better than average ☐	excellent ☒
DRIVE -STABILITY	Has poorly defined goals ☐	Appears to set goals too low ☐	Appears to have average goals ☐	Appears to strive hard to achieve. ☒	Appears to set high goals ☐
OVERALL	unsatisfactory ☐	Substandard ☐	Average ☐	above average ☐	Outstanding ☒

Form No. 22. *Continued*

Page two. PROSPECTIVE EMPLOYEE INTERVIEW REPORT FORM.

This applicant: ☒ Should be hired ☐ Should not be hired If no, please state reasons:

Type of work that you feel the applicant is most qualified to do:

OUTSTANDING CANDIDATE FOR FUTURE SUPERVISORY POSITION

Additional Comments: THIS EMPLOYEE IS THE TYPE WE WOULD LIKE TO GROOM FOR MONTHLY SALARIED POSITION.

Signature of Interviewer Ronald D. Macklin

Job Descriptions

Once the employee is hired, you will want him to know exactly what his responsibilities are going to be. A brief job description can be very valuable. Make sure the employee knows the limits of his authority and the proper chain of command within your organization. Many times the job description can be combined with your accident prevention efforts so that you have a checklist for each responsibility.

Figure 30 shows a practical example of an accident prevention and job training checklist for a loader operator. Notice how the checklist incorporates the best of both programs.

1. Point out and demonstrate the use of all loader controls. Cover the service and maintenance requirements of the equipment and establish an inspection and service schedule. Warn against slips and falls and their relationship to good housekeeping in and around the equipment.
2. Explain the responsibilities of the loader operator and the need for him to have an ongoing useful communication with the chaser, knot bumper, skidder operator or yarder engineer and the log truck drivers.
3. Review all logging safety codes applicable to his function. Put emphasis on the fact that the loader operator is responsible for the condition and stability of all logs loaded by him, both at the landing and on the truck during transport. It is his job to warn drivers of hazards and to be sure that they comply with all applicable safety requirements while on the landing.
4. Explain the hazards associated with moving the machine from one location to another and the need for firm roadbeds. If the loader is rubber tired, go over the proper procedures for road travel: excessive speed, steep grades, sharp curves and the need for ground guidance and snubbing in hazardous situations.
5. Go over the landing layout, building proper decks and sorting requirements. Point out dangers of pinch points and requirements for minimum 3-foot counterweight clearance. The storage or placement of items on the loader should not be allowed. No one should approach or be around the loader unless the operator says it is safe to do so. He must be ever aware of the position of the chaser or other persons on the landing. He must at no time suspend a log or equipment over a person.
6. Explain the communication procedure between the operator and the yarder engineer, the chaser and others while clearing the landing shoot. Point out the dangers of YUM and PUM being thrown by the loader or allowing material to roll or slide off the landing, endangering the crew working below.
7. Review standard and alternate yarding whistle signals and the signals used to signal and move log trucks and equipment on the landing. A copy of the whistle signals should be given to the operator to keep with him in the loader.
8. Review the proper procedure for building proper loads and the dangers of unsaddled logs, peakers, trying to shift loads, handling oversized logs and damages to trucks caused by rough handling of logs.
9. Explain proper trailer unloading and the need for extreme care while the chaser or driver is in the loading area. If log branding is being done after the log is on the truck, special caution must be taken and he must be sure the load is stable and the person branding is in the clear before loading or moving another log.
10. If using the front-end loader, the operator must use grapple arms, move slowly and be aware of unstable or uneven ground. If knot bumpers are working on the landing, he must be aware of their location at all times and be sure they are well in the clear. He must also control the movement of tractors and skidders on the landing and arrange the loading area so that the landing personnel and skidder operators are not subjected to hazards.
11. Explain the travel capability of the machine and the braking capability.
12. On a regular basis, review all aspects of job performance and provide additional training if such is required.

A checklist like this can be made up for all the jobs in your operation. It will help the employee and it will help your company.

Figure 30. Accident prevention and job training checklist: loader operator

Form No. 23. New employee indoctrination checklist

NEW EMPLOYEE INDOCTRINATION CHECKLIST   Date 4/23/82

Employee Name RONALD P. WEBER # 502

Position or Job HOOKTENDER

Will Report To RON MACKLIN

Starting Salary #11.60 Hours of Work FIRST SHIFT

Vehicle Assigned NONE Credit Cards NONE

Insurance Eligibility Date 5/23/82 Coverage ALL PROGRAMS

Actual Starting Date 4/23/82 Vacation Eligibility 4/23/83

✓ Special Equipment Issued

✓ Employee Handbook Issued

✓ Other Company Policies Covered

✓ Covered Safety Orientation Check-List

✓ Explain use of Company Vehicles

✓ Explain use of Company Credit Cards

✓ Equal Opportunity Policy Explained

✓ Use of Time Cards Explained

✓ Rigging Check-Out Procedure Explained

✓ Use of Fuel Pumps Explained

✓ Certification Verified

✓ Employee Number Assigned

✓ Insurance Booklets Issued and Explained

✓ Safety Manual Issued and Explained

EMPLOYEE STATEMENT:

"I certify that all the above information was covered and that I completely understand all the items pertaining to my employment, safety, and conduct."

Ronald P. Weber
Employee Signature

Carmen Macklin
Indoctrination Given By

NEW EMPLOYEE INDOCTRINATION

Now that the necessary steps have been taken in the initial hiring process, you are ready to indoctrinate the new employee into the mainstream of your company work plan.

Now is the time for the employee handbook or the company policy manual to serve its ultimate purpose. If you have both, the handbook can serve as a summary of all the details covered in the policy manual. Form No. 23 is a checklist that can be used during indoctrination so that everything necessary will be covered.

POLICIES, FRINGE BENEFITS AND EMPLOYEE INCENTIVES

Usually, the first thing a new employee wants to know is what benefits he has and how they are administered by the company.

We have mentioned that one of the most difficult management tasks is getting and keeping good employees. Many things make up what makes an employee happy and what keeps him happy over a period of time. Part of this puzzle is found in the area of employee benefits and incentives. These benefits have been estimated at as much as 25% of the total cost of labor. The average cost of benefits is 15% for all employers.

Benefits are an expanding cost area for all employers as employees seek increased and nontaxable benefits in addition to increased salaries. Keeping people happy is expensive.

Salaries and Raises

Salaries and wages will likely be one of your greatest headaches. How do you best determine how much a job is worth and when to give an employee more money?

Let's talk about salaries first. How do you determine what a job is worth? The best place to start is the marketplace. What is everyone else paying for the same worker?

One of the best ways to find out what other companies are paying is to contact them and ask. It sounds a little brazen but it works. Call up other businesses like yours and ask them if they would like to participate in a salary comparison survey. Explain that you are not trying to stir up competition among the businesses for good employees. What you are trying to do is to arrive at a fair price for a job and to make it the same price everyone else is paying. You don't want to be the highest or the lowest paying firm.

When I did this, most of the companies I contacted were very interested in seeing the survey. All I really wanted was the average salary for a particular job. Most of the companies gave me their lowest and highest salaries for each of the jobs. I prepared a spread sheet showing each company and how it compared with the rest of the companies participating in the survey. It really worked out well and it stabilized the wages in the area for months at a time.

My goal was to be somewhere in the middle of the salary ranges. You will find that is the best place to be, even when the job market is slow.

With that information, I established a minimum, middle and top range for each job classification. When I started out a green employee, I started him out at the minimum for his job. If the employee was highly experienced and I could see that he was going to be a definite asset, I started him out nearer the top of the salary range. I made it a practice never to start a new employee at the maximum salary for his job classification. If you do, you find yourself boxed in later when it is time for a salary increase and there are no promotions available. Leave the employee room to grow within his job description.

For the sake of discussion, let's take a choker setting job; call the top class A, the middle class B and the bottom class C. Here is how the salary range looked at the time I did this:

Class A	$9.50 per hour (maximum hourly)
Class B	$7.00 per hour
Class C	$5.50 per hour (entry level)

You will notice the $4.00-per-hour spread between the entry level and the maximum for this hourly job. This gives an employee some room to grow. This eliminates many salary problems and unhappiness between employees of the same job description.

Once an employee reaches the top of his salary range, he is usually ready to be promoted to the next level job. If there are no openings, you can create an incentive plan to allow him to grow even beyond the maximum for his job. We will explain the use of this plan later in the chapter.

Raises must be planned and much thought should be given to the method you use in handing them out. A policy should be established as soon as possible, if you do not already have one. A system of regular reviews should be worked out. This system can be as simple or as complicated as you wish. Documentation should be used in this review process, as a record for the employee and to protect you as the employer.

Employee Evaluations

A system of regular evaluations is an important part of keeping employees happy. Employees have a right to know how you feel about their performance. If they are doing a good job, they should know it. If they are doing a bad job, they should know that too. It is difficult for an employee to change his behavior or job performance if he does not know he is doing something wrong.

Form No. 24 is a fairly uncomplicated evaluation sheet that you might be able to use. The forms aren't absolutely necessary, but they can be helpful. They provide a record of everything that is covered and they can be retained in the employee's file for later reference. You will notice that almost every possible area is listed on the evaluation form. The areas that do not apply to a particular employee can be left blank.

It is good to evaluate an employee at least every 6 months. Many times it is possible to have a salary review at the same time.

Form No. 24. Employee evaluation form

EMPLOYEE EVALUATION FORM Date 1/5/82

Employees Name Larry Lampkin

Employee # 321 Job Title Yarder operator

PURPOSE

This evaluation form is used to take a personal inventory, to pinpoint weaknesses and strengths and to outline and reach agreement on a practical improvement program for each employee. It is good if the evaluations can be done at least twice each year.

Time in this position 4½ years Time with current supervisor 3 yrs

THE POSITION The employee and immediate Supervisor should briefly summarize the important duties and responsibilities of the position.

Responsible for daily operation of yarder. Concerned with the safe operation of the equipment. Does periodic inspection of cables, drums, brakes & other vital parts of yarder. Helps supervise set up & break down of yarder when relocating or re-positioning.

ACCOMPLISHMENTS DURING EVALUATION PERIOD List and discuss specific major accomplishments of the Employee during the past evaluation period.

Employee is consistently responsible for high production & safe operation of his yarder. Keeps equipment running & reports major difficulties to his supervisor as they occur.

Form No. 24. *Continued*

Page two. EMPLOYEE EVALUATION FORM.

STRENGTHS OF EMPLOYEE The immediate Supervisor should list the strengths of the Employee and discuss these with the Employee. Cite specific examples wherever possible.

HARD WORKER, AS EVIDENCED BY HIS HIGH PRODUCTION
GOOD ATTENDANCE – VERY DEPENDABLE

AREAS FOR IMPROVEMENT The Supervisor should list the areas where he feels improvement on the part of the Employee is needed and discuss these with the Employee. Cite specific examples wherever possible.

EMPLOYEE SHOULD FINISH THE CERTIFICATION
PROGRAM FOR YARDER OPERATORS

PERFORMANCE PLAN After discussion, the Supervisor and Employee should jointly prepare a performance plan for the next twelve months. Cite specific work assignments with steps designed to improve the Employee's personal development.

EMPLOYEE WILL FINISH THE CERTIFICATION PROGRAM
ON FEBRUARY 5TH.

Larry Lamphier 1/5/82 Ron Macklin 1/5/82

Employee's Signature | Date | Supervisor's Signature | Date

Your company policy regarding evaluations could read something like this:

Employee Evaluations

1. The company has installed and will maintain a job evaluation system for all workers (with the exception of those noted herein), in an effort to apply sound principles of measurement to determine what each job in the organization is worth and to gauge the effectiveness of each employee.
2. In maintaining the system of job evaluation it will be the policy of the company:
    a. To write into the description of each job all duties that the incumbent will regularly be called upon to do, so that there will be a minimum number of occasions when he will be required to do work outside his job description.

       NOTE: It must be understood that management has the right to divide the duties in the manner it considers most efficient, and will not ordinarily include a seldom required duty that raises the job grade. If such a duty is regularly required, it must be included in the job write-up. Job grades must be based strictly on job descriptions.
    b. To promptly write up and evaluate jobs as provided elsewhere in these rules and policies, whenever new jobs are created, or the contents of the old jobs are significantly changed.

Bonuses and Gifts

Incentives such as bonuses and gifts are usually governed by your profit picture and your generosity as an employer. Your company policy should make it very clear that such things are solely at the discretion of company management and that they will not always be an annual affair.

One of the main things to remember about bonuses is to make sure that the proper withholdings are taken at the time the bonuses are issued. This practice will not only keep you in line with Uncle Sam, but it will eliminate any problems for you and the employee regarding his payroll withholdings at the year's end. It will all be taken care of in the proper way and at the proper time.

As the boss, you should let your bookkeeper or accountant determine how and when to issue bonuses and gifts. Gifts can be handled a little differently than a bonus in terms of taxes, so use your staff expertise in this area.

Vacations

Vacations will usually be one of your more expensive fringe benefits, averaging 8½% of the labor costs for executives, 7½% for salaried employees and about 6½% for hourly employees.

It is strictly up to management how liberal the company vacation policy will be and what the eligibility requirements will entail.

The policy should be clearly written and could appear as shown in Figure 31. Form No. 25 is a sample vacation request and approval form.

Holidays

Many small logging firms have no provision for paid holidays. If you decide to use this particular benefit, keep the following important points in mind:

It is the policy of ABC Company to grant vacations with pay to provide qualifying employees with periods of rest and recreation in recognition of services performed.

Administrative guidelines

A. Vacation allowances

1. The vacation allowance for eligible weekly payroll employees for each vacation year is based on length of service according to the following schedule:

Completed Years of Service	Vacation Allowance
First year	1 week
Second to fourth years	2 weeks
Fifth to tenth years	3 weeks
Tenth or more years	4 weeks

The weekly vacation allowance is the normally scheduled work week.

2. Bi-weekly and monthly payroll employees (same schedule as above).
3. Employees are not eligible for paid vacation during the first year of employment. One full year must be completed before paid vacation can be taken.

B. Eligibility provisions—General

1. Vacation benefits under this policy apply only to permanent full-time employees of this company.
2. The vacation year is the 12-month period commencing with the employee's first employment anniversary date and each subsequent anniversary date.

C. Eligibility provisions—Holidays, leave of absence, illness occurring during the employee's vacation.

1. If a designated holiday falls during an employee's vacation, the employee will be eligible for additional time off equal to the holiday time, with pay.
2. If an employee takes a leave of absence during the year, his vacation time for that year will be cut 1/12th for each full month of leave granted. If the employee leave is less than one full month, his vacation will not be cut. An employee entering active military service or commencing a leave of 6 months will receive vacation pay for unused earned vacation and accrued vacation since the employee's most recent anniversary date.
3. An employee will not receive additional time off due to illness or disability occurring while on vacation.

D. Vacation pay

1. Weekly vacation pay will be computed by multiplying the employee's current straight-time hourly rate by the number of hours of vacation allowance, i.e., \$5.00 per hour × 40 hours = \$200.00. Bi-weekly pay is number of weeks vacation × weekly pay rate.
2. Vacation pay will be paid on the last scheduled work day prior to the vacation, provided the vacation has been prearranged with the office and management.

E. Vacation scheduling

1. Vacations will be scheduled by management, with consideration being given to staffing requirements, employee's seniority and employee's preferences, in the order listed.
2. Vacation may be scheduled at any time during the employee's vacation year and may be taken over the vacation year end, provided it commences in the vacation year in which it is due.

Figure 31. Vacation policy—ABC Company

3. Vacations should normally be taken in units of at least one week. An employee may, however, take one week of his vacation as separate days subject to management approval.
4. Whenever possible, employees should take their total vacations. Pay in lieu of vacation will not be granted unless absolutely necesssary to meet staffing requirements, and only if approved by management.

F. Eligibility provisions—terminating employees
1. In all cases, a terminating employee will receive pay in lieu of any unused vacation allowance earned as of the employee's most recent employment anniversary date.
2. In addition, a terminating employee will receive pay in lieu of vacation allowance accrued during the current vacation year, provided the termination was due to one of the following reasons:
   a. Resignation with notice.
   b. Release.
   c. Retirement.
   d. Death of the employee.

G. Annual company shutdown
1. The company reserves the right to close the entire operation for 2 weeks, during summer shutdowns or winter layoffs, and require all eligible employees to use their earned vacation at that time.

Figure 31. *Continued*

- Your costs due to absenteeism before and after holidays will be increased. This problem can be eliminated by paying for the holiday only if the employee works the day before and the day following the holiday. If the holiday is in conjunction with a weekend, then it would be the last working day prior to the holiday and the first working day following the weekend and holiday.
- Clearly outline all your procedures and specific holiday dates in your procedures manual and employee handbook.
- If a holiday falls on Sunday, Monday will be the paid holiday.
- If a holiday falls on Saturday, Friday will be the paid holiday.
- The most common paid holidays for U.S. and Canadian companies are:

New Year's Day	Independence Day
Washington's Birthday	Labor Day
Victoria Day (Canada)	Veterans Day
Memorial Day	Thanksgiving Day
St. Jean Baptiste Day (Quebec, Canada)	Christmas Day
Dominion Day (Canada)	Boxing Day (Canada)

You must decide which holidays to use and how many to use. Many small logging companies cannot afford paid holidays of any kind.

- Follow the guidelines used in making up the vacation policy. Try to cover all contingencies and eliminate confusion.

Form No. 25. Vacation notification and approval form

VACATION NOTIFICATION AND APPROVAL FORM

Date 3/18/82

To: ALFRED SMITH Employee # 65
(Employee's Name)

In accordance with our Company Policy, as of 3/20/82, you have been in our employ for 10 year(s) and are entitled to 2 week(s) paid vacation.

To assist us in scheduling vacations, please indicate your first, second, and third choice for vacation time below and return both copies of this notification to your supervisor by 5/1/82. One copy will be returned to you indicating approved vacation time.

	First Choice		Second Choice		Third Choice	
	Begin*	End	Begin	End	Begin	End
1st week	8/2/82	8/13/82	8/16/82	8/27/82	8/30/82	9/10/82
2nd week	"	"	"	"	"	"
3rd week						
4th week						
5th week						

Date 5/10/82

We are happy to approve your vacation time for the following period:

Period of AUGUST 2, 1982 THRU AUGUST 13, 1982

Signature of Approving Official [signature]

* Since normal vacation weeks start on Monday, please use Monday dates.

Sick Leave

Some small logging firms have this benefit only for their monthly and executive people. If you use this benefit, you might want to consider the following things:

- Have a predetermined allowable number of sick days per year, usually, one day per month, not to exceed 5 to 10 days per year. Some larger firms allow accumulation of sick days from year to year, but most do not.
- You can determine to pay sick leave only on provision of a doctor's certificate.
- Some companies allow no pay for the first one to four consecutive days, followed by full pay for a predetermined time. This method is usually used in

connection with a group insurance policy so that the only cost to your company is the insurance premium. Employee income is protected during a long-term illness, yet single "sick days" are less frequent when the employee is not paid.

- Clearly outline all your procedures regarding sick leave in your employee handbook or policy manual.

Insurance: Medical, Hospitalization, Life, Dental, Optical and Other

Most small logging companies cannot afford to participate in insurance plans unless they go through an association or some other group plan. There are many plans available.

If you are seriously considering this kind of benefit for your employees, it would be worthwhile to get several package presentations from different insurance carriers. The carrier will provide all the forms and instruction booklets you will need to administer such a program.

Your carrier will usually recommend a set procedure for qualifying your employees. The carrier will usually want a probationary period for each new employee and will greatly appreciate the information you have regarding each employee's medical record.

The insurance carrier will help you write a procedure for administering the insurance program. The extent of the program will depend on the amount of money you wish to pay out for such coverage. Many small companies pay only the employee's premium; the employee then pays the premium for his own dependents. This takes much of the financial burden from your shoulders.

Retirement Plans

Again, most small loggers cannot afford to have retirement plans. As for insurance, there are association and other group plans available. The loggers who have retirement plans are usually very selective about whom they let participate. If you decide to have such a plan, it is good to have rigorous eligibility requirements. For example, an employee may become eligible for the plan only after a predetermined period of time. This could be as long as 5 years or as little as 1 year. You could also specify a predetermined age group that could participate.

Consult your CPA or your attorney before setting up such a plan. There are new laws governing retirement plans and the way the federal government will monitor them during their existence.

Severance Pay

Severance pay is a benefit not commonly used anymore in small companies, especially logging. The turnover ratio is usually far too high to make such a practice feasible. You could reserve the practice for executive or monthly employees. The final decision is yours, but be very careful if you decide to use this benefit. Again, consult your CPA and your attorney.

A. Establishing expense accounts and company credit cards.
   1. Reason for establishing.
      a. Credit cards are being issued to recognize key employees for performance and achievement.
      b. Expense accounts will allow staff to be reimbursed for company expenses.
B. Expense account and credit card guidelines.
   1. The following guidelines will govern the use of all company-provided credit cards and expense accounts.
      a. The cards are to be used for company-related travel and entertainment expenses or supplies. The following items are acceptable in this category:
         1) Airline tickets.
         2) Bus tickets.
         3) Train tickets.
         4) Cruise tickets.
         5) Hotels and motels.
         6) Meals.
         7) Car rentals.
         8) Gasoline and other products for company vehicles or business.
         9) Shows and athletic events involving company clients.
         10) Golf and other sports involving company clients and business contacts.
         11) Miscellaneous supplies.
      b. The use of employee's cash or personal credit cards for company-related travel and entertainment expenses or company supplies. The same items are acceptable as shown in items 1 through 11.
C. Establishing expense account limits
   1. Your expense account monthly limits are determined by company policy.
      a. Limits established for you should not be exceeded.
      b. Each employee's limits will be established by management.
      c. Your monthly limits will be monitored by the office.
D. Issuance of cards
   1. Cards will be issued by the office.
      a. Upon termination or layoff, all cards must be returned to the office on the last working day of the employee.
      b. Lost or stolen cards must be reported to the office immediately.
      c. Damaged cards should be turned in at the office and new ones issued.
E. Submittal of charge slips and receipts
   1. All charge slips are to be turned in at the office on a weekly basis for company credit cards.
   2. All charge slips or receipts from your personal accounts or cash purchases should be turned in monthly with careful documentation.
      a. Make sure all slips contain the date, location, name of item purchased and person or persons involved.
      b. If entertaining customers, be sure to include their names and company affiliation; also list what was discussed during the meal or event.
      c. Please do not accumulate receipts and charge tickets over a long period of time. They must be accounted for at the end of each month.
      d. Careful documentation cannot be stressed enough. The IRS can be quite difficult if records are not complete regarding entertainment and travel.

Figure 32. The use of company credit cards and expense accounts

Company-Furnished Automobiles and Driving Expenses

It is a common business practice to provide vehicles to key personnel within a company and to provide crew buses to transport working personnel to the job sides. It is also a common practice to reimburse employees for using their own vehicles on company errands. A tough system of controls and procedures should be provided to eliminate employee abuse of the system and to circumvent certain tax write-off problems. Tough procedures should be instituted to govern where the vehicles go and who takes them.

Here are some common abuses of company cars:

- Driving the car home each evening and back to work the next morning. This could cost you a fortune if the employee lives a long way from work. The best way to work this is to establish a maximum distance that the cars or trucks can be taken at night and on weekends. Any miles over that maximum would be reimbursed by the employee to the company at so much per mile. Even easier yet would be not to allow the vehicles to be driven home at all. Make the employee drive his own car to work and take his own car home.

  There is, however, one case where it would be better to allow your employee to take the rig home. If the job site is closer to his home than it is to your shop, it would save you money to let him take it home.
- Close controls should be maintained over the gasoline and other materials used in the vehicles.
- When you pay mileage for an employee to use his own car for company business, you should require trip mileage to be turned in on the employee's expense account. Make sure you do not exceed the maximum payment for mileage set down by the IRS. Make sure your policy clearly states that the mileage to and from work in the employee's car is not covered by your reimbursement plan.

Expense Accounts

During the course of your company's growth, you may determine that it would be a good idea to allow some of your key people the use of company credit cards and expense accounts. This is a widely accepted business practice, but you should be very careful to clearly outline the use of such privileges and the proper gathering of documentation for all legal business expenses.

The outline in Figure 32 can be used when making up your company policy governing this employee incentive. When everything is clearly outlined, there can be no misunderstandings in regard to the use of company credit cards and expense accounts. Make sure your office staff keeps a handle on the use at all times. Make sure you consult your CPA regarding the use of company credit cards; he will offer assistance in the record-keeping department.

Suggestion System

Your own employees can be an outstanding source of innovative and money-saving ideas. In order to get these ideas, it is beneficial to start a simple suggestion sys-

Form No. 26. Employee suggestion form

EMPLOYEE SUGGESTION FORM — Date 3/14/82

PLEASE WRITE YOUR SUGGESTION CLEARLY, INDICATING EXACTLY WHAT IS TO BE DONE; IF YOU NEED MORE SPACE OR NEED TO DRAW A SKETCH USE THE BACK OF THIS FORM OR ATTACH A SEPARATE SHEET OF PLAIN PAPER. PLEASE PRINT OR TYPE CLEARLY.

My Suggestion is: THAT ONLY SPECIFIED INDIVIDUALS BE ABLE TO CHECK OUT RIGGING FROM THE RIGGING SHACK. THE WAY IT IS NOW, ANYONE CAN GET THE KEY & CHECK OUT RIGGING. MANY ARE FORGETTING TO WRITE DOWN WHAT THEY ARE TAKING & THE RIGGING INVENTORY SHEET IS NOT ACCURATE.

My Suggestion will accomplish the following things:

1. ONLY THOSE RESPONSIBLE WOULD CHECK OUT RIGGING.
2. THESE INDIVIDUALS WILL REMEMBER TO FILL OUT THE CHECK-OUT SHEET.
3. INVENTORIES WILL BE ACCURATE.
4. LOST RIGGING WILL BE HELD TO A MINIMUM.

Employee Name Ronald R. Weber — Employee # 502

Address 8608 N. 31ST DR. SPRINGFIELD 97477 — Telephone 746-8995

ALL SUGGESTIONS BECOME PROPERTY OF THE COMPANY TO USE AS IT SEES FIT.

tem. A simple form such as Form No. 26 can be used in such a system.

If each suggestion submitted can be recognized by management in some way, it makes the system much more effective. If a suggestion is used, the employee should be rewarded in some way.

When a suggestion is received by management, the employee should receive a notation to this effect. A simple card can be used for this. Figure 33 shows an example of such a card.

When it is determined that a suggestion can be used, a value can be placed on it at that time. If the suggestion develops a large savings, a percentage of that savings for the first year could be passed on to the employee. If the suggestion results in a minor savings or results in safer working conditions or some other direct but intangible benefit, a standard cash award can be made or a premium gift of some kind can be given.

It is good perhaps once a year to recognize the outstanding suggestor or suggestion for that period. A certificate or similar gift is often appropriate.

The program can be as simple or as complex as you want it to be. Your company does not have to be large to benefit; small companies need good ideas too, and this kind of a program will get them.

Your suggestion system policy might read like the one shown in Figure 34.

Overtime

Many times the use of overtime is necessary to complete a job on time or to deliver a product when it is needed. The use of overtime should be carefully controlled by company management.

The use of overtime is only worthwhile when it accomplishes the desired result. Much money can be wasted if overtime is not properly controlled.

The graph in Figure 35 shows what a typical employee's productivity and alertness look like during a day's work. You will see that at the end of the day your people are going to be far less productive than at the beginning of the day. It is

```
YOUR SUGGESTION HAS BEEN
     RECEIVED
________________________

A careful evaluation will be
made to determine the feasi-
bility of your idea.

You will be notified when a
decision has been reached &
a determination will be made
at that time.

THANK YOU.
```

Figure 33. Suggestion receipt notification

The more efficient the company is, the easier our work and the more profitable for all of us. Suggestions for the improvement of operations or procedures may be submitted by nonmanagement employees to supervisors for evaluation. Suggestions should be detailed, organized, and show money, time or material savings.

The suggestion, if endorsed, will be forwarded to the company president for review. If a suggestion is approved and implemented, the employee who devised the suggestion or plan will receive a cash reward not to exceed $________. All suggestions will be acknowledged in writing, and will include an explanation of any action to be taken by the company.

We believe that our employees are capable of making constructive and valuable suggestions and that they should be rewarded well for their ideas. For this reason we have instituted a suggestion system.

All employees except those who are in executive, managerial, supervisory, technical or administrative positions are eligible to participate in the system.

Suggestions that contribute to the safety of employees will result in the payment of a $________ award.

$________ will be paid for a suggestion that leads to an improvement in the way a job is accomplished, but for which no cost savings can be determined.

________% of the first year's net saving will be paid for a suggestion that results in a savings to the company in the cost of delivering its products.

Forms on which suggestions can be written out and submitted to management are located in various designated areas and in the office.

All suggestions will be acknowledged.

Please note:

1. The company maintains an open mind regarding the improvement of its operating procedures, and encourages its employees to make suggestions to this end. Suggestions are welcomed on such subjects as safety and ways to save labor, money, energy, time and materials.
2. All suggestions should be submitted in writing. Management will be happy to discuss any suggestion with its author.
3. Cash bonuses are given to the writer of adopted suggestions.

Figure 34. Suggestion system policy

reasonable to assume from this graph that the quality of work will be less and the danger of accident much higher after normal working hours. This important factor should be kept in mind when scheduling overtime.

Form No. 27, Overtime Request and Approval, can be used as a permanent record of all overtime authorized and the reasons for such overtime. A company overtime policy should include the information given in Figure 36.

Company Parties and Gatherings

Party or social activities are strictly up to management discretion. If you enjoy this sort of thing, sometimes it's better than incentives or gifts.

If your company adopts this form of benefit, the following things should be kept in mind:

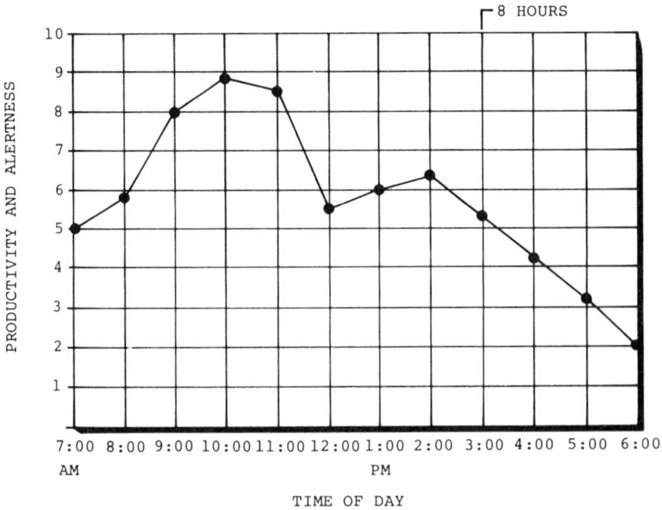

Figure 35. Productivity and alertness of a typical employee

Form No. 27. Overtime request and approval form

OVERTIME REQUEST AND APPROVAL FORM

Date 5/8/81

Employee Name SID SOUTHWARD

Employee # ____________ Shift FIRST

Request that the above named employee be permitted to work 9 hours on 5/8 & 5/11/81.

☒ Overtime ◯ Make-up Time

Employee will work on Job No. 12

Describe WILL BE DECKING LOGS FOR LOADING

Reason requested LOGS MUST BE READY AT 5:00 AM EACH MORNING

Request prepared by [signature]

Request Approved by [signature]

Denied by ____________

Reason: ____________

Definition

Overtime is the time in excess of 40 hours which is actually worked in a 1-week pay period (day shift Monday through night shift on Sunday).

Circumstances in which overtime may be authorized

A. Overtime may be authorized only when it is not possible to adjust staffing patterns and/or employee work schedules to provide essential staffing in the following types of circumstances:
   1. Emergency circumstances
      a. These are adverse circumstances of a temporary nature which reduce normal staffing or increase staffing requirements. For example:
         1) The unexpected absence of an employee whose position must be filled to avoid a disruption of necessary work;
         2) The necessity to immediately perform an unexpected heavier-than-normal volume of work to avoid disruption of necessary services;
         3) Situations which are uncontrollable, such as heavy snowstorms, or other interference.
   2. Nonemergency circumstances
      a. These are situations in which overtime work is considered to be the best way to make necessary staff available to handle a particular situation. For example:
         1) To provide essential staffing where vacancies exist due to budgetary problems or to the unavailability of qualified job applicants;
         2) To perform an unusually heavy volume of work of a temporary nature in the implementation of an approved job;
         3) To perform work that cannot be performed during normal working hours without disrupting necessary activities or services.

B. To avoid overtime, management, whenever possible, is expected to grant compensatory time off during the same pay period for time worked in excess of that normally scheduled.

Authorization of overtime work

Only approved supervisory personnel may authorize overtime work in emergency circumstances. In nonemergency situations, overtime work must be authorized in advance by upper management personnel. In requesting such authorization, the manager concerned must clearly justify that overtime constitutes the best means of meeting the situation.

Payment for overtime work

1. Employees will receive 1½ times their regular rates of pay for each full quarter-hour worked in excess of 40 hours in a weekly payroll period.
2. Only time in excess of 40 hours actually worked in a pay period will be considered in determining eligibility for overtime payment of 1½ times the employee's regular rate. This includes break periods and time worked while "on call" but ex-

Figure 36. Overtime policy

cludes time spent "on call" but not working, meal periods, sick leave, vacation, leave of absence, and so forth.

3. During the week in which an employee receives holiday pay, the number of holiday hours which the employee is paid will be considered as hours "actually worked" in determining eligibility for overtime payment for such week.
4. Overtime pay will be calculated on the basis of the employee's regular rate of pay inclusive of shift premiums, on-call premiums, and so forth. Shift premiums will be paid for overtime hours worked in accordance with policies governing shift premiums.

Distribution of overtime among employees

1. To the extent possible, overtime will be distributed equally among full-time employees of the same classification in a particular job who are willing to work overtime, provided the employees concerned are equally capable of performing the duties required in the judgment of the supervisor or manager.
2. To the extent consistent with No. 1 above, overtime work will be offered to full-time employees considered equally capable of its performance in the order of their length of continuous employment.
3. An employee who refuses overtime will, for the purpose of equal distribution, be charged with the number of hours he would have worked had he agreed to work and will not be asked to work overtime again until his turn comes again, or an emergency situation clearly exists in which he is the most logical person to work the overtime required.

Procedure

Overtime hours actually worked are to be recorded on the weekly time card and must be approved by the management personnel in charge.

Weekly payroll time reports must be sent to the office each week, where they will become a permanent record for 3 years. Daily time reports must be maintained in the office for a minimum of 3 years.

Every effort should be made to adhere to this policy and all employees should be made aware of its existence upon joining the company.

Figure 36. *Continued*

- Establish an annual budget for this activity so that funds can be allocated in advance.
- Have one of your management staff in charge of the entire program. Have them outline to you what is to take place prior to the actual scheduling of the facilities and the expenditure of company funds—unless you like surprises. If you do, just let them go and see what happens later.
- Have the arrangements made so that control of the event is maintained. If alcohol is served, it is good to have a limit so that the party doesn't get out of hand.
- Remember that your company is responsible for everything that happens as a result of your gathering.

Form No. 28. Personal fringe benefit report

PERSONAL FRINGE BENEFIT REPORT

Name Ronald P. Weber

Emp. # 502 Date 12/30/82

Listed below are the values of your Company Fringe Benefits for the year to date:

Vacation Pay	$ 0
Holiday Pay	$ 500.00
Medical Insurance	$ 560.00
Life Insurance	$ 320.00
Sick Leave	$ 0
Company Vehicle	$ 1000.00
Profit Sharing	$ 350.00
Retirement Plan	$ 0
Other... Bonus	$ 500.00
Total Benefits Paid	$ 3230.00

- Remind your staff member in charge to get bids from several places that have the proper facilities to serve you. If it is a local affair, the likelihood of competition in bidding is even better.

Fringe Benefit Visibility

Your company can get maximum value for its expenditures by making sure your employees regularly know the value of their benefits package.

This can be achieved by using a Personal Fringe Benefit Report like Form No. 28. Such reports may be issued periodically (usually on a quarterly basis) and can be included with the employee's paycheck and withholding summary.

As you can see, this shows the employee the exact value of his fringe benefits and how much it is costing the company to provide them. It is a good employee relations tool.

SAFETY AND TRAINING PROGRAMS

As another employee benefit and to comply with state and other laws regarding safety, a formal program should be instituted and given the complete support of you, the owner, and the management of your company.

A Safety and Training Director should be appointed for your company. His duties should include the following things:

- Formulating, administering and making necessary changes in the accident prevention program of the company.
- Submitting directly to the Safety Committee regular monthly, weekly or daily reports on the status of the safety program.
- Acting in an advisory capacity on all matters pertaining to safety as required for the guidance of the management, the general manager and all supervisory personnel.
- Maintaining the accident record system and making necessary reports for OSHA and other agencies. Securing supervisors' accident reports and checking corrective action taken by supervisors to eliminate accident causes.
- Supervising the safety training of all employees.
- Correlating safety work with the work of the personnel people to include proper selection and placement of employees.
- Making personal inspections and supervising inspections by his staff and by the Safety Committee for the purpose of discovering and correcting unsafe conditions or unsafe work practices before they cause accidents.
- Maintaining professional outside contacts to exchange information with others and to keep the program up to date.
- Supervising fire-prevention and fire-fighting activities where they are not the responsibilities of other departments within the company.

The Safety Committee

The function of the Safety Committee will be to review existing policy and safety regulations so as to guarantee the company's compliance with all existing laws and regulations regarding logging safety. The committee will also review questions and suggestions offered by employees to help make the work environment safer.

The Safety Committee will meet monthly at a time convenient for all members. The committee will consist of six members to be appointed by the director.

The committee meetings will be recorded and all pertinent data will be passed on to the employees through a safety memo. Each employee will receive a copy of the memo for his own safety manual.

The Safety Committee will be responsible for the investigation of all accidents involving serious or potentially serious injuries. Reports of each investigation will be written and discussed at the Safety Committee Meetings.

The Safety Committee will be totally responsible for:

- Development and application of safety standards for facilities, equipment, tools and work methods, using applicable, legal and voluntary codes, rules and standards as a minimum.
- Safety inspections to identify potential hazards.
- Accident investigation to determine the cause of accidents and prevent the recurrence of such accidents.
- Accident records and accident cause analysis to determine accident trends and provide targets for corrective action.

- Education and training in general safety principles and techniques. On-the-job safety instructions by the supervisor and periodic supervisory contacts with new instructions, follow-up and general safety motivation.
- Protective equipment to provide personnel protection in hazardous areas.
- Industrial hygiene studies to identify potential health hazards and development of safe protective measures.
- Safety publicity and promotion to step up program interest and participation.
- Off-the-job accident prevention and cooperation with public and private agencies to promote the application of accident prevention principles to non-work activities.

Expressions of management interest include enforcement of the policy and program, recognition of good safety records, review of safety reports and participation in meetings, safety banquets and other events that show support for the company safety efforts.

Management must take the lead in keeping interest alive by:

- Emphasizing that production and safety go together.
- Setting a good example. For instance, if company rules require employees to wear ear and eye protection while doing certain work, management should observe these same rules and regulations when visiting such work areas. If the boss does not wear protective equipment, why should the employees?

Use of Unsafe Equipment or Unsafe Work Practice Report

As a follow-up to the supervisors' responsibility to continually stress safety and safe work habits, you should use the Unsafe Equipment or Unsafe Work Practice Report (Form No. 29). This report must be completed any time an unsafe piece of equipment is reported or discovered or you find an employee using unsafe work practices.

There are three copies in the Form No. 29 set. In the case of unsafe equipment, one copy goes to the shop, one copy to the office for follow-up and you retain one copy. In the case of unsafe work practices, one copy goes to the offending employee, one copy to the office for follow-up and you retain one copy.

The form should be filled out in the following way:

1. Check which report you are filing.
2. Date report is completed.
3. Appropriate job name or number.
4. List type of equipment.
5. Company number on vehicle.
6. Give clear description of problem—clutch slipping, bad brakes, and so forth.
7. If reporting unsafe work practice, name of employee.
8. Which warning: first time? second? or third?
9. Nature of violation—horseplay, lack of attention, and so forth.
10. Action taken—corrective action such as scheduling additional training. If equipment, what are the repairs recommended or action needed to solve the problem?

Form No. 29. Unsafe equipment/unsafe work practice report

UNSAFE EQUIPMENT REPORT ☐ (1)
UNSAFE WORK PRACTICE REPORT ☒ (Check One)
Date 5/4/81 (2)
Job # 12 (3)
Equipment Type (4)
Company # (5)
Brief Description of Problem (6)

Name of Employee in Violation PIO MARTINEZ (7)
Nature of Violation (9) HORSE PLAY ON THE LANDING
WARNING: (8)
First ☒
Second ○
Third ○

Action taken to rectify situation or recommended repairs
(10) EMPLOYEE WARNED ABOUT HORSE PLAY & THE DANGERS INVOLVED

Other Information: (11) EMPLOYEE VERY COOPERATIVE, GOOD ATTITUDE, WILL IMPROVE THIS AREA

Prepared by: [signature] (12)

11. Other information. Anything pertinent to the matter not listed elsewhere on the form.
12. Sign report after checking for completeness.

Make distribution of copies as described earlier.

TERMINATION OF EMPLOYEES

"The ultimate personnel function" it's been called by men wiser than we. Regardless of what you call it, no one likes to say "You're fired!"

Unfortunately there are times when there is no other alternative except to let an employee go. When this unpleasant task becomes necessary, it is good to have a company procedure for getting it done. Usually this is the only way to weed out an undesirable employee or an employee who cannot fit into your preconceived mold or pattern of behavior.

Termination of an employee should never be based on emotion or on a spur-of-the-moment problem. Termination should come only as a last resort, after everything has been tried to salvage the person. Termination should not be used as an immediate solution, unless circumstances warrant such action. In cases of theft, violation of company policy on numerous occasions or other serious matters, immediate action has to be taken. If it's a matter of personality conflict or other compatibility problem, there are better solutions. If an employee is truly a constant problem, he or she will usually prove to be so over a long period of time. There is not much use in trying to salvage a constant headache.

Form No. 30. Employee exit interview form

EMPLOYEE EXIT INTERVIEW FORM — Date 2/14/82

Name JEFFREY DOE — Employee # 321

Social Security # 513-30-6433 — Job Title FMC OPERATOR

Supervisor's Name MACKLIN — Date Hired 7/14/79

Date Separated 2/14/82 — TYPE OF SEPARATION: Retirement ☐ Layoff ☐

Resignation ☒ Discharge ☐ Other ☐ __________

REASON OR REASONS FOR SEPARATION: (Please check all that apply)

RESIGNATION	DISCHARGE	LAYOFF
Physical Condition ☐	Lack of Ability ☐	Temporary Work ☐
Family ☐	Bad Personality ☐	Reduction in Force ☐
Continuing Education ☐	Lack of Drive ☐	Other ☐
Better Job Offer ☒	Lack of Efficiency ☐	
Going into Own Business ☐	Lack of Cooperation ☐	
Disliked:	Dishonesty ☐	
Hours ☐	Rules Violation ☐	
Supervisor ☐	Excessive Absence ☐	RETIREMENT
Type of Work ☐	Excessive Tardiness ☐	Age ☐
Wages ☒	Accident Prone ☐	Medical ☐
Working Conditions ☐	Other ☐	
Other ☐		

COMPLETE WHEN THE EMPLOYEE IS DISCHARGED:

When was employee notified? __________ How was employee notified? __________ When should final check be issued?

COMPLETE IF EMPLOYEE RESIGNED:

New Employer LOWELL BUTTE LOGGING — Location LOWELL, OR.

Nature of Work FMC OPERATOR — Pay 9.50 — Hours SAME

You should develop a system for warning employees and documenting infractions of company policy. If you are going to let someone go, you must have the ammunition to get the job done. If a union is involved, your work will be much harder. If no union, your job is somewhat easier, but you should still arm yourself with the necessary facts to defend your stand.

As we mentioned earlier, you might offer severance pay if you wish to do so. If possible, you should give the person ample notice of their departure. Losing a job is a traumatic experience in itself; losing it on the spur of the moment can be a devastating experience.

If the employee is resigning or quitting on his own, we always hope that he or she will give adequate notice. Two weeks is the accepted rule-of-thumb. If the employee has an executive position that is hard to fill, it would be good if he could give you more than the usual amount of time to find and train a replacement.

Form No. 30. *Continued*

Page two. EXIT INTERVIEW FORM

COMPLETE IF EMPLOYEE IS LAID OFF:

Was employee offered a transfer to another position?__________

What job was the employee offered?__________

Why did the employee refuse the transfer?__________

IN ORDER TO DETERMINE THE REASON FOR AN EMPLOYEE RESIGNING THE FOLLOWING QUESTIONS SHOULD BE ASKED:

What did you like about your job? WORKING YEAR ROUND

What did you dislike about your job? THE PAY WAS TOO LOW

How did you feel about your compensation? TOO LONG BETWEEN RAISES

How did you feel about your supervision? OK

How did you feel about our training program? OK

Did you feel you were being used in the right job? OK

How do you feel about the company? NO HARD FEELINGS

Would you stay if some changes were made? YES

What changes would you expect? MORE MONEY

What is the real reason for your leaving? MONEY

COMMENTS OF INTERVIEWER:

ECONOMIC CONDITIONS DO NOT ALLOW US TO RAISE THE EMPLOYEES SALARY.

Jeff Doe — Employee signature

Carmen L. Miskimin — Interviewer signature

Final action COULD NOT RAISE SALARY, EMPLOYEE RESIGNED.

Regardless of the situation, the termination process should be made as smooth and as painless as possible. It is good for you and it is good for the employee. Form No. 30 is an excellent Exit Interview Form.

SUMMARY

People are our most valuable asset. Retaining good people requires much work and expense.

The programs outlined here help to keep your people satisfied. Use them where you can and select those that your people will value the most.

People require maintenance, just like roads. But, like roads, they can make your journey through life much smoother.

Remember: *accomplishment comes through people.*

CHAPTER 10

The Equipment Part of Your Business

Next to your people comes your equipment. Without the proper equipment and people, logging would be one of the most impossible jobs in the world.

In this chapter we will cover some important areas concerning equipment. What you as the boss need to know about it, how you keep track of it at all times, how you schedule field and in-shop maintenance, repair and replacement, whether to leave it dirty or keep it clean and many other things important to your success in the logging business.

MAINTENANCE PHILOSOPHY

How do you as the owner feel about your equipment and its use? For example, some companies buy equipment, use it, abuse it, tear it up and junk it out for parts. Many times we have seen equipment junked out long before its useful life had been reached, just because the crew didn't care about the equipment and its proper upkeep.

We have all seen equipment on the road and in the woods that looked like an accident going someplace to happen. The sad part is that it usually does happen and someone is hurt or killed because of badly maintained equipment.

Your maintenance philosophy has a lot to do with which way your equipment is heading and whether people will look at your equipment and say, "Old Macklin's stuff is going to hurt someone yet. Have you ever seen such a bunch of junk?"

We hope you are one of the owners or managers who cares about your equipment. If you care about it, you will never be at the same disadvantage as those who do not.

KEEPING TRACK OF YOUR EQUIPMENT

If your company is small enough, keeping track of your equipment will never be a problem. You will be with your equipment each day and you will always know where it is and what shape it's in. But what if you have loaders, 8 log trucks, 5 high-lead towers, 6 Cats, dump trucks, fire trucks, fuel trucks, FMCs, rubber-tired skidders, 10 crummies, 20 pickups and so on? How are you going to know where everything is and who is responsible for its use and abuse? Here are some helpful procedures you can use.

Equipment Inventory Procedure

Equipment inventory listing is important for the following reasons:

- It provides you with a complete listing of your company's assets.
- It provides a way of controlling and knowing the location of your equipment at any given time.
- It provides the list that you need for the county each year in regard to your personal property taxes.
- It provides your accountant the list he needs to prepare your depreciation schedule each year.

Form No. 31 is an Equipment Inventory List. The list should be updated annually with a special notation for equipment that was sold or traded. Make sure to keep accurate records of where the equipment went and when it was sold or traded. The state agencies want to know what you do with every piece of equipment and you must be able to tell them.

As new items are acquired, they are just added to the bottom of the list as your company continues to grow from year to year.

Form No. 31. Equipment inventory list

EQUIPMENT INVENTORY LIST

Purchase Date	Equipment Description	Qty.	Serial No.	Company No.	Alpha Code	Mkt. Value
4/15/79	FMC - TRACKED SKIDDER	1	0123	28	FMC	25,000
4/16/79	KENWORTH 1975	1	4567	29	KW	24,500
4/18/79	KENWORTH 1979	1	8901	30	KW	42,500
5/4/79	SKAGIT TOWER BU-80T-80	1	87568	31	SKT	95,000
5/16/79	LINKBELT LS98	1	3210	32	LL	60,000
6/18/79	HOMELITE POWER SAW	3	04738 57643 04739	33	PS	1,650
8/18/81	MOBIL RADIOS	3	FA14763 FA14764 FA14765	34	MR	4500
3/15/81	SKAGIT TOWER BU-80 T-80	1	SN37568	35	SKT	120,000
5/14/81	MARION LOADER	1	M-5687	36	ML	18,000
6/18/81	BOOSTER	1	758643	37	PUP	1,600
8/19/81	PEERLESS LOWBOY	1	PL46857	38	LB	17,200
(1)	(2)	(3)	(4)	(5)	(6)	(7)

Looking now at Form No. 31, notice how it is completed:

1. The purchase date or date you took delivery.
2. A description of the equipment, brand name, and so forth.
3. Quantity column is for smaller items, like typewriters.
4. Serial number of equipment.
5. Company number assigned by your inventory control.
6. The alphabetical code used to identify the equipment.
7. The fair market value at the time of purchase.

We have found it best not to reuse numbers when equipment is sold or traded. It is less confusing simply to retire the numbers as the equipment is marked off the list.

This same list can be used in conjunction with your insurance program. Your insurance company will require a list like this each time your policy comes up for renewal. This should have been item 5 in our earlier list of important reasons for keeping an inventory list.

An Equipment Numbering System

A company equipment numbering system can be very valuable when used in conjunction with the Equipment Inventory List.

In Form No. 31 you will remember that column 5 provided a place for the company number.

The numbering system can be as simple as 1, 2, 3, 4 or as complicated as listing the month and year of purchase as well as a number and description code. In Figure 37 you will see an example of the coded equipment numbering system.

The following is an example of an equipment description code that we used in our logging business:

BL = Bantam Loader
CAR = Automobile
CAT = Dozer
CBR = Citizen Band Radio
CK = Choker
DT = Drip Torch
DT = Dump Truck
FBT = Flat-bed Truck
FEL = Front-end Loader
FL = Fork Lift
FMC = Tracked Skidder
FT = Fire Truck
GR = Grader
GT = Gas Truck
KW = Kenworth
LB = Low Boy
LL = Linkbelt Loader
LTR = Log Trailer
ML = Marion Loader
MR = Mobile Radio
MT = Mini Tower
OE = Office Equip.
PC = Pump Can
PS = Power Saw
PU = Pickup
PUP = Booster
RG = Rigging
RS = Rubber-tired Skidder
SE = Shop Equip.
SHR = Space Heater
SKT = Skagit Tower
SLY = Slackline
ST = Shop Truck
SUB = Suburban
TBT = Thunderbird Tower
TPL = Tree Puller
VAN = Van
WRP = Wire Rope
WTR = Water Truck

It is a simple coding system and can be added to as you acquire new equipment. It is very helpful in developing a filing system for your equipment records.

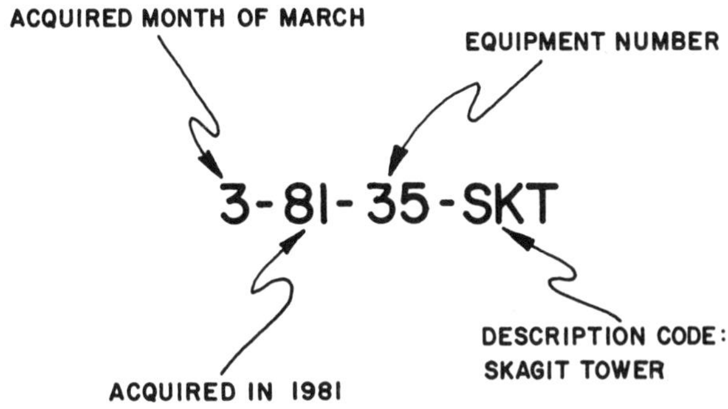

Figure 37. Equipment numbering system

The advantages of using a more complicated system are evident in the details the number code can reveal. If you have a lot of equipment, use this system. If you have just a few pieces of equipment, stick with the plain number system.

Equipment Files

As we just mentioned, equipment files are very important. They give you a place to put everything having to do with each piece of equipment in the way of documentation and records.

A folder should be prepared for each piece of equipment. Figure 38 shows how the folder tab should be filled out. The folder tabs can be color-coded to show the status of the equipment. Here is a sample color coding:

Green = Equipment is free and clear of all financial encumbrances.
Yellow = Equipment is currently under a purchase contract.
Blue = Equipment is currently under a lease contract.
Red = Equipment is paid for but is currently being used as collateral on a loan.

Figure 38. Equipment folder

Each of the file folders should contain the following information:

- The Equipment Record and Description Form.
- All contracts and purchase papers.
- Title (if such is in your possession).
- Maintenance history.
  Shop work orders.
  Record of expenses.
- All other documentation not mentioned above.

Records and Descriptions

Once the equipment folders are prepared, the Equipment Record and Description Form (Form No. 32) should be prepared. Notice how the form is to be used and the extent of the detail it contains:

1. The manufacturer (Kenworth, etc.).
2. Model number.
3. Serial number (from body or block).
4. Date machine was purchased.
5. Company number.
6. Insurance carrier's name and policy number.
7. Is the item being depreciated or not? Type of depreciation if it is on the schedule.
8. Current year.
9. Amount of depreciation this year (if any).
10. Current market value.
11. Total miles driven this year.
12. Total hours operated this year.
13. Gallons of fuel consumed this year.
14. Total purchase price (including interest).
15. Cost to replace at this time.
16. Color of equipment.
17. Cubic inches or horsepower.
18. Number of axles.
19. Gross weight.
20. Current license number.
21. License expiration date.
22. Cost of license.
23. Dimensions of equipment (if known).
24. Place of purchase.
25. Employee assigned to operate.
26. Gross weight capacity.
27. CB radio serial number.
28. Mobile radio serial number.
29. Who purchased if equipment was sold.
30. Price equipment sold for.

Form No. 32. Equipment record and description form

EQUIPMENT RECORD AND DESCRIPTION FORM

Make KENWORTH (1) Model CONV. (2) Serial Number 8901 (3)

(4) Date Purchased 4/18/79 Control Number/Asset Number 4-79-30-KW (5)

(6) Insurance Coverage MUTUAL ACME, INC. Depreciation (7)

Year/Schedule	19 79 (8)	19 80	19 81	
Depreciation	(9)			
Current Value	(10)			
Miles Driven	(11)			
Hours Operated	(12)			
Fuel Consumed	(13)			

(14) Total Purchase Price 42,500.00 Replacement Cost 56,000. (15) Color BLK & YELL (16)

(17) Engine Size 400 CUMMINS Axles 3 (18) Gross Wt. 58,00 (19) License # T3758 (20)

Expiration Date 5/1/81 (21) License Fee $45.00 (22) Size STD. (23)

Purchased From HANDY'S KENWORTH (24)

Assigned To LEONARD SMITH (25) Load Capacity 78,000 (26)

RADIO: CB Serial # 3756843 (27) Mobile Comm. FA-2374567 (28)

Equipment Sold To (29) Selling Price (30)

Capital Gain (31) Loss (32) Date Sold (33)

Conditions of Sale (34)

Mileage When Sold (35) Fuel Consumed: Weekly (36) (Gal) Monthly (37) (Gal)

Notes:

(38)

Prepared By: [signature] (39)

31. Capital gain (for accounting purposes).
32. (Loss) (for accounting purposes).
33. Date equipment was sold.
34. Conditions of sale (i.e., cash, contract, etc.).
35. Final mileage when sold.
36. Average weekly fuel consumption.
37. Average monthly fuel consumption.
38. Explanatory notes or additional comments.
39. Prepared by.

This completed form will always stay in the file folder for that particular piece of equipment.

THEFT AND VANDALISM OF EQUIPMENT

In recent years, theft and vandalism has become a major problem for businesses using heavy equipment. Nationally, well over $400 million worth of equipment was stolen in 1980. Vandalism accounts for millions of additional dollars lost. What can a logger do to help fight the problem?

Here is a field security checklist that provides some basic ideas that might help with the problem in your area:

____ Contact local authorities upon job start-up and make them aware of the types of equipment that are popular items for theft or vandalism.

____ Maintain an ID list of serial numbers of all equipment on the job; provide local authorities with a copy of this list.

____ Do not allow equipment to be moved off the job without your approval; verify all "office" orders that unexpectedly send equipment to other jobs. (Phony work orders are easy to make.)

____ Prime theft times are early morning at start-up and in the evening at quitting time; keep your people aware of theft potential at these times.

____ Be aware of people "hanging around" a jobsite. Professionals usually case a job first. Be suspicious; approach unknown individuals and ask what their interest in the job is. Write down vehicle license numbers of questionable parties.

____ Painting your equipment special or wild colors won't stop or even slow down a professional thief—they will simply repaint it on the spot or immediately after taking it.

____ Know your own people. Some thieves rely on inside information. Check out the places your people frequent after work hours—these are contact points where information, either willingly or unknowingly, exchanges hands.

____ Don't expect things like electronic ignitions to stop thieves—the pros call it child's play.

____ Situate larger machines around smaller ones—the more a thief has to move, the less likely he is to hit your side.

____ Sonic alarms and electronic "beeper bugs" (transmitters that send out a traceable signal) are good deterrents and recovery aids.

____ Lock the drivetrain, if at all possible, on rolling stock. Professional thieves say it's too hard to beat or takes too much time to overcome.

____ Don't worry about fences or barriers; the thief will drive right over them.

____ Remember that security dogs can be put out of action with a piece of drugged meat.

____ Watchpersons can help prevent theft and vandalism.

____ Report missing equipment to your local law-enforcement agencies immediately; time becomes a benefit to the thief the longer that the theft remains unreported.

Most law-enforcement agencies will work with you in setting up a program of this type.

EQUIPMENT PERFORMANCE AND EVALUATION

By using the forms developed for getting information from the woods to the office, you will be able to evaluate the overall performance of your equipment. Here is a list of the forms in the Timber Harvesting System and what they will provide you in the way of information:

Form No. 2: Helps you to know what is being moved, how long it takes to move in various kinds of weather and where it's all located.
Form No. 3: Lets you know the productivity of your yarders in turns per day and logs per day.
Form No. 4: Provides a daily record of all productive time, downtime, fuel consumption and lubricants. This allows you to watch for non-productive time and keep track of breakdowns and minor repair work done by the operator.
Form No. 6: Provides a monthly summary of all the daily inputs from Form No. 4.
Form No. 9: Shows the number of logs loaded daily by the loader.
Form No. 11: Gives a complete report for each log truck—mileage, fuel and trip destination and distance.
Form Nos. 14 and 15: Provides fuel consumption and mileage on all vehicles not covered by other forms.

All the forms combine to provide you with a total report on:

- Machine transport frequency and cost.
- Location of machines.
- Performance and productivity of machines (hours operated versus product moved).
- Reliability of machines (downtime).
- Repair requirements.
- Fuel and lubricant consumption.
- Mileage comparison.
- Distance factors.
- Destinations.
- Effect of weather on equipment productivity.
- Effect of other random variables on equipment.

When evaluating all this information, the following variables will determine the overall performance of your equipment:

- Stand density.
- Slope.
- Terrain.

- Soil conditions.
- Undergrowth.
- Size and weight of trees.
- Distance to landing.
- Distance to mill.
- Weather conditions.
- Personnel operating and supporting equipment.
- Waiting time on landing and at the mill.
- Loading and unloading time.
- Road conditions.

As you can see, most of these items are beyond your immediate control. But they must be dealt with on every job. Never forget to include them in your bid preparation and job planning. They can kill you on your logging prices if you fail to include them.

TYPES OF MAINTENANCE

In a discussion of maintenance we must relate to the complete support activity of the entire equipment function. To accomplish this we must first cover the six major maintenance types.

Field Maintenance

Maintenance in the field can be the backbone of your service program. It is the first line of defense against breakdowns. Much of the work can be accomplished by your woods personnel without having to send your mechanic out to the side. Here is a list of things your operators can do on their own:

- Proper tension adjustments.
- Ground level lubrication.
- Fluid level checks.
- Routine pressure checks.
- Routine wear checks.
- Anti-freeze checks.
- Battery checks.
- Filter checks and replacement.

Anything more complex should be handled by an experienced mechanic. A record of these routine items can be kept on Form No. 4, the Equipment Use Form, entries 13 and 14.

Routine and Preventive Maintenance

Routine maintenance must be planned and scheduled if equipment breakdown and interruptions in productive time are to be minimized. Included in this category are repair work, ranging from routine adjustments of a mechanism to rebuilding an entire unit, as well as cleaning and upkeep.

Corrective Maintenance

Corrective maintenance must be carried out when a piece of equipment repeatedly breaks down for a given reason and it is determined that the breakdowns can be eliminated by adjusting, replacing or modifying some part of the equipment. A major overhaul of the equipment might be needed, but usually this is much more likely to involve a small adjustment or change, easily made on the job site.

Periodic Maintenance Overhaul

Periodic maintenance overhauls are an essential function of your maintenance people unless the more specialized pieces of equipment are sent to the manufacturer for attention. It is often more expensive, but there are certain pieces of equipment that must be handled in this way. Many times schools are offered for a minimal fee to instruct your own mechanics in some of this sophisticated equipment. It is wise to send them.

Turnaround Maintenance

Turnaround maintenance is a vital function in the logging industry. Many times it is absolutely vital, while the unit is still in service, that all maintenance work be preplanned and organized to a point where the unit can be overhauled or maintained at the lowest possible cost in the shortest possible time when shut down. All parts are ordered in advance and everything is ready in the shop before the equipment is brought in for maintenance.

Breakdown Maintenance

Breakdown maintenance is of course the basic responsibility of your maintenance people and, as such, has to have priority so equipment can be put back into service as quickly as possible. In addition, your maintenance people should ask, "Why did the unit break down?" Investigation should be made into the cause of the trouble in order to avoid a similar breakdown in the future.

Logging management is becoming more and more aware that maintenance includes aspects other than dollar costs or savings. Of course, maintenance costs must be kept to a minimum, but equipment utilization figures, downtime hours, safety of personnel and the general appearance of the equipment are also of vital importance. Our ultimate goal is optimum use of every maintenance dollar spent.

DETERMINING THE EXTENT OF MAINTENANCE AND REPAIRS

Your maintenance philosophy will dictate some guidelines in this area.

In some organizations the shop foreman makes the decision on routine maintenance and repairs up to a $1,000 maximum. At this point you may wish to enter into the decision-making process. During this discussion it is decided whether to purchase new or used parts and whether to do the work in your own shop or send it to the manufacturer's local representative shop. During this time you might also decide whether to repair the equipment at all. Perhaps you might want to rebuild the unit completely or replace it with a new piece of equipment.

If this is the case, a major allocation of funds is necessary and your controller should be notified in advance of your decision.

Reasons for Renovating

- To ensure the best efficiency, productivity and safety.
- To correct faults that could lead to major equipment damage or injury to company personnel.
- To ensure maximum trade-in or resale value.

Reasons for Rebuilding

- Extra life warrants spending the money. (Sometimes 50% of the replacement cost can be justified.)
- New equipment costs too much.
- We know what the machine has been through and trust that we can make it like new again.
- Newer models won't fill our needs as well.

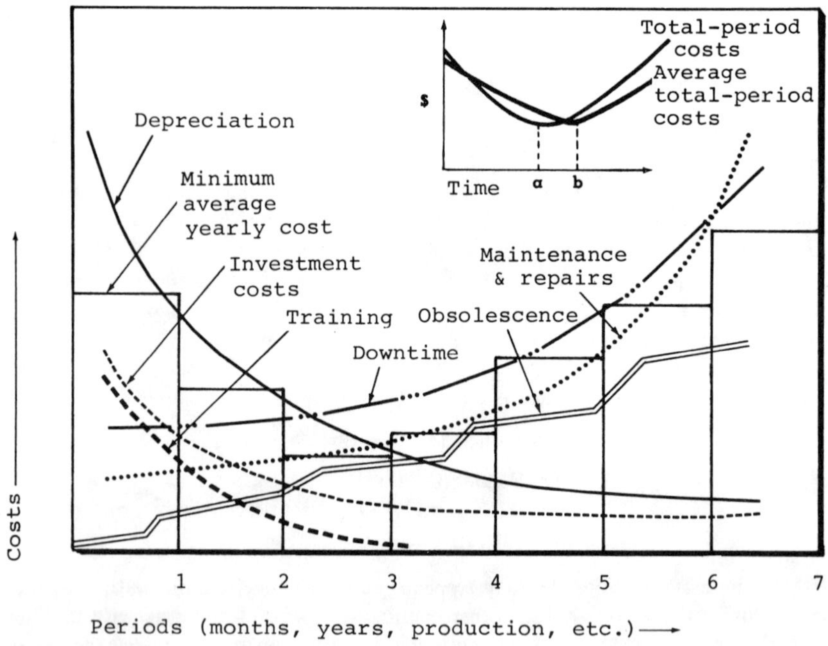

Figure 39. Charting your equipment costs

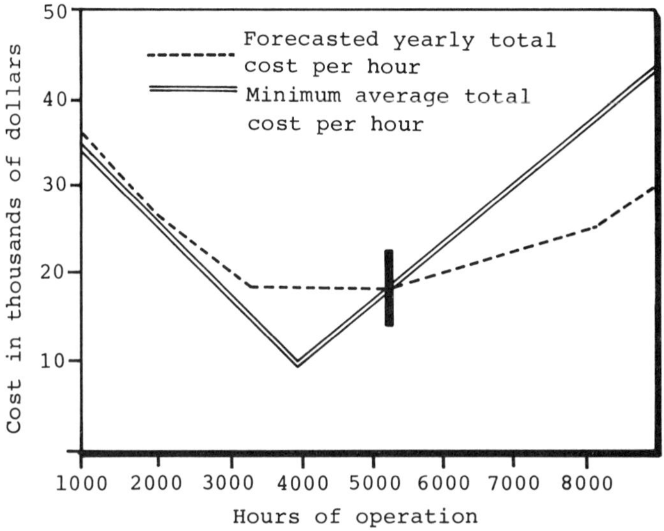

Figure 40. Comparing forecasted with actual equipment costs

Reasons for Replacing

- Not enough potential life left in the old machine.
- Equipment exists that is more suitable to our needs.
- The machine has reached the absolute end of its scheduled life.
- We cannot properly correct the problems in the existing machine.
- We can afford to invest in newer equipment at this time.

When productivity and efficiency continually fall, it is time to replace. Each machine generates the expenses shown in Figure 39. By monitoring them, you can see how much individual units are costing your company. Owning costs decline in the first year's use, while we see a climb in operating costs. When related to production, every machine will reach a point of diminishing return on investment. The small inset in Figure 39 shows the curve averages, with point "b" being a good time to replace the unit. In Figure 40 you can see that by establishing a profile of your equipment's operating and owning expenses, you can better judge the effects of rebuilding versus replacing. When the forecasted and actual costs become the same, it is time to look for a new piece of equipment.

LEAVE IT DIRTY OR KEEP IT CLEAN?

Is it worthwhile to spend money on appearance as well as function? You will find that one has a lot to do with the other in the long haul.

The cost of cleaning equipment on a regular basis is very small compared with the benefits it can produce.

Clean equipment provides three very desirable things:

- It increases the payload in your trucks.
- It makes for a good public image.
- It increases the productive life of the machine.

Experts in this area have presented the following logic:

- Buildup of unwanted weight, such as slush, snow, ice, mud, tar, oil and grime, can add as much as 1,500 pounds of extra weight to a piece of equipment.
- Equipment of any kind can only take so much filth before it gets into trouble. The cleaner it is, the easier it is to locate a defect and get it repaired. Did you every try to find a splinter in your finger when it was covered with grease? Your equipment is the same.
- Clean engines and underparts can save up to 25% of your mechanic's time during inspection, routine maintenance and repairs.
- Mechanics can work much faster and will receive fewer injuries on clean equipment.
- Your public image will be improved when people see that your equipment is clean and well-kept.
- Your operators and drivers will take pride in a shiny piece of equipment. They will take better care of what they operate when it looks good.

The investment in a high-pressure combination cleaner and high-potency detergent is well worth it.

LOWERING MAINTENANCE COSTS

Maintenance costs can be reduced in several ways. How many mechanics do you have? If you are like most companies, you have only one or two permanent people. What do you do when they are swamped? Do you hire another man? Why add another man to the payroll when you can subcontract the extra work? Have a contract mechanic come in on an hourly basis and eliminate all those fringe costs associated with a company employee. When the extra work is gone, so is the extra mechanic. No layoff, no hurt feelings.

Do you keep a parts inventory? If you have storage space, you can sometimes make large purchases to take advantage of periodic quantity discounts. Also, if you know a particular item is going up in price tomorrow, buy an extra today and put on the shelf.

Try to use reconditioned or used parts where practical rather than new ones. The machine doesn't know the difference. You might have some parts fabricated locally rather than purchase them. Many times you can get a better part for the same or less money. Even if it costs you more, you usually get better materials and workmanship.

Sometimes it is not what you know but who you know that can really save you maintenance money.

EQUIPMENT MAINTENANCE RECORDS AND SCHEDULES

Once it has been determined that a piece of equipment must be repaired or that it is time for routine maintenance, the Preventive Maintenance and Repair Work Order Form (Form No. 33) can be used.

Form No. 33. Preventive maintenance and repair work order form

PREVENTIVE MAINTENANCE AND REPAIR WORK ORDER FORM

Date 7/12/81

(2) Equipment No. 4-79-30-KW Make and Model KENWORTH CONV. (3)

(4) Year 79 Mileage 32380 (5) License No. T3758 (6) Time In 4:00 P.M. (7)

(8) Promised out 7/15/81 Regular P/M [X] Special Need [ ] Emergency [ ] (9)

PARTS USED (Attach Invoices) Mechanic Assigned ELMER (10)

Quantity	Description	Price
3	FILTERS	36.00
4'	HYDRAULIC HOSE (12)	3.50
4	BULBS	2.80
(11)		(13)

Work Accomplished: (14) REGULAR PREVENTIVE MAINT.

(15) Work Done: In Shop [X] By Vendor [ ] (Please attach all invoices)

Service performed by vendor: (16)

Total Parts Cost: $42.30 (17)

Total Vendor Cost: 0 (18)

Total Hours Worked in-House: 3 @ 15.00 (19)

Job # Where Equipment was last Working 12 (21)

Signature of Mechanic: Raymond Brauer (22)

TOTAL JOB COST

$ 87.30 (20)

This form was developed to keep track of maintenance and repair expenses on each piece of equipment. During the course of work, all invoices for parts, special labor and other items relative to the job should be attached to this work order. The form is completed in the following manner:

1. Date equipment enters the shop.
2. Equipment number assigned.
3. Make and model of equipment.
4. Year of manufacture (or acquisition).
5. Mileage or odometer reading.
6. License or registration number.
7. Time in (a.m. or p.m.).
8. Date or time promised out.
9. Check if regularly scheduled preventive maintenance.
10. Mechanic's name.
11. Number of parts used.
12. Description of parts used.
13. Price of parts used. (Attach all invoices for outside purchases. If part is from inventory, specify.)
14. Brief description of work done. (If more room is needed for entry, use back of form.)
15. Where was work done? In shop or by subcontractor? (If by vendor, attach all invoices.)
16. If vendor, briefly describe work done.
17. Total parts cost. (Inventory parts will be priced by office.)
18. Total vendor cost (services only).
19. Total hours worked in house on this work order.
20. Total job cost (will be completed in the office).
21. Job number where equipment was last working.
22. Signature of mechanic.

These forms will not be filed into job folders. They will be filed in the equipment folders explained in Chapter 9.

SUMMARY

To get the job done, you must have the equipment. To complete the job, your equipment must continue to operate properly on a daily basis. To operate properly, the equipment must be cared for.

If you as the boss develop a good maintenance philosophy, you will have reliable equipment.

Remember: *productivity through properly used equipment.*

CHAPTER 11

Maintenance of the Entire System

Once you have established the control system, there are certain things you must do to keep it going and to keep it healthy.

MAINTAIN ABSOLUTE CONTROL

As the boss, you must maintain control of every aspect of your overall business. Periodically check to see that all the forms in the system (that you chose to use) are being completed by all employees and that your office staff is making the appropriate entries and maintaining the reporting system. The system functions best as a whole. If any part of it dies, it will adversely affect the remaining parts. If you don't support it 100%, it will die and all your initial work will have been in vain.

REEVALUATE AND BENEFIT FROM YOUR MISTAKES

If a particular portion of the system fails to provide the data you need, change it to better serve your purpose. The system serves as a foundation on which to build your company. Keep adding to the system and tailoring it to meet your individual needs more closely. It is a place to begin and it is a place to end.

If you discover, through the use of the system, that you have made many mistakes in conducting your business, don't be discouraged. Reevaluate your actions, make the necessary changes and benefit from your past mistakes.

MOVE AHEAD TO GREATER PROFITS

There is no reason why you should not be able to make a profit, if you can keep working without being shut down due to the economy or other variables.

Obviously, we are all victims of the economy and its periodic ups and downs. But if we are prepared in advance to face setbacks, we will survive.

With the control system you will know, from day to day, the status of your company and its financial performance. You will never be forced to make a major decision without all the information you need.

Remember what a famous man once said: "To get profit without risk, experience without danger and reward without work is as impossible as it is to live without being born."

This famous man must have had a logger in mind.

ABOUT THE AUTHOR

Ronald R. Macklin understands business and logging. The principles and practices described in this book result from more than 20 years of management and administrative experience, including 5 years in the logging industry.

Macklin began developing a business management system for loggers in 1977, when he became secretary-treasurer for Carter's Inc., an Oregon logging company. His responsibilities included overseeing all financial records, fiscal planning, purchasing and employee training programs. The forms and procedures he designed received "real world" testing in this company with a 36-million-board-foot gross volume.

Macklin expanded his experience in the logging industry while director of membership and special projects for the Associated Oregon Loggers. There he structured a leasing company, developed an evaluation of the benefits of leasing versus buying and set up a seminar on simple bookkeeping practices. He also developed forms and forms control systems for the association and was involved in funding for safety training and research programs.

Now Macklin runs his own consulting business, Triangle M Company, in Lowell, Oregon. As a consultant and in staff positions with Motorola, Macklin has prepared and designed training manuals and audiovisual presentations for such companies as Weyerhaeuser, Xerox and Boeing. He has more than 800 hours of business-oriented classroom teaching experience, as well as experience in time-study and work-factor analysis.

Author of dozens of articles and booklets, Macklin's publications in the logging field have included a regular column for The *Log*, AOL's magazine, articles on maintenance and repair costs for *Timber/West* and question-and-answer columns for *Logging Management*.

His professional memberships include the American Management Association, the American Society of Notaries and the International Association of Business Communicators.